Advanced Series in Agricultural Sciences 6

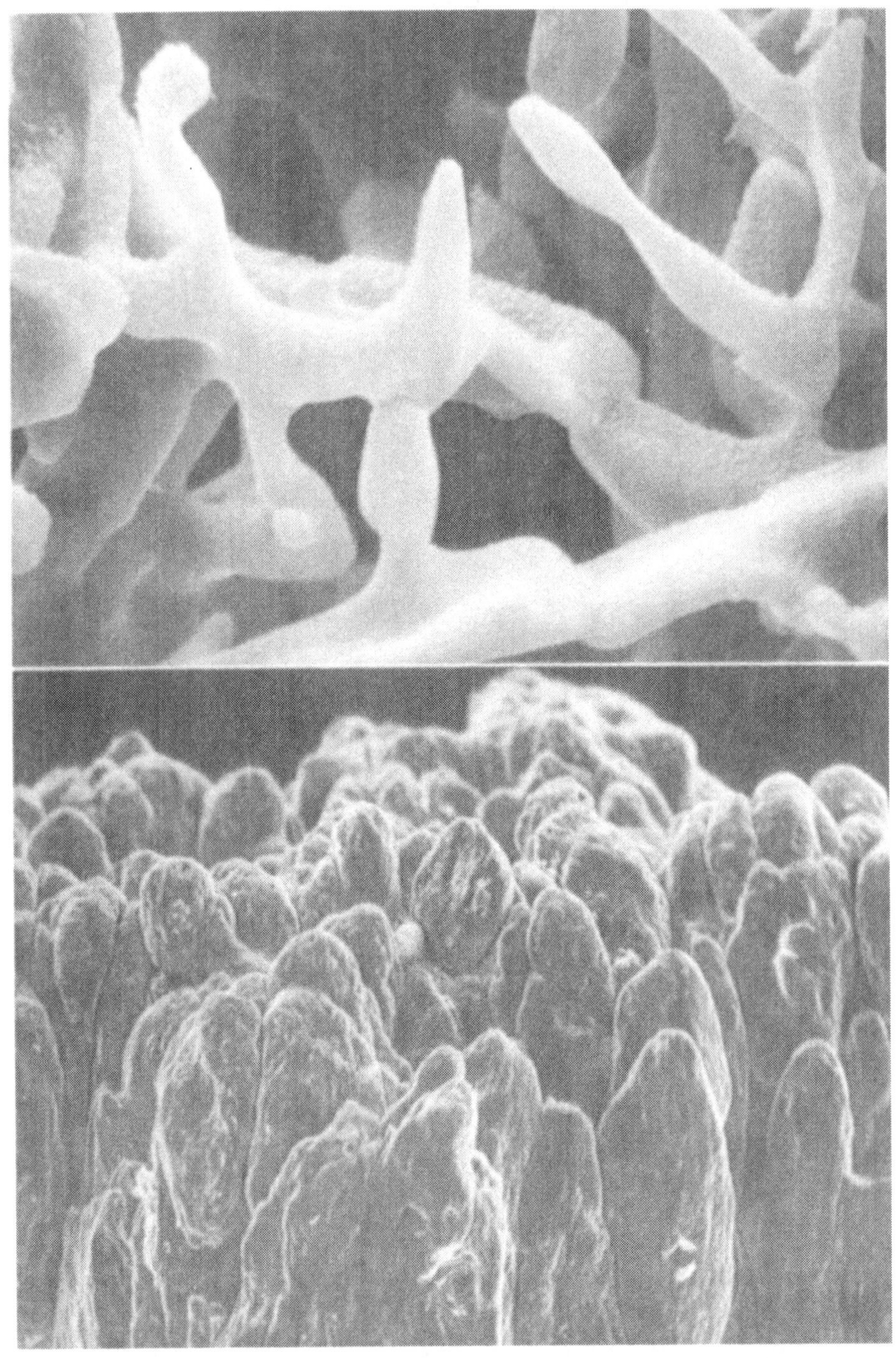

Precambian fossil plants. Top: *Witwateromyces conidiophorus* (× 5000). Bottom: *Thuchomyces lichenoides* (× 70). (Courtesy of Dr. D. K. Hallbauer)

J. E. Vanderplank

Genetic and Molecular Basis of Plant Pathogenesis

Springer-Verlag
Berlin Heidelberg New York 1978

J. E. VANDERPLANK, Ph. D., D. Sc., D. Sc. (Agric.), honoris causa
Department of Agricultural Technical Services
Plant Protection Research Institute
Agricultural Buildings, Beatrix Street
Private Bag X 134, Pretoria 0001, South Africa

With 3 Figures

ISBN-13: 978-3-642-66967-5 e-ISBN-13: 978-3-642-66965-1
DOI: 10.1007/978-3-642-66965-1

Library of Congress Cataloging in Publication Data. Vanderplank, J. E. Genetic and molecular basis of plant pathogenesis. (Advanced series in agricultural sciences; 6) Bibliography: p. Includes index 1. Plant diseases—Genetic aspects. 2. Plants—Disease and pest resistance—Genetic aspects. 3. Plant diseases. 4. Plant proteins. I. Title. II. Series. SB 731.V 26. 632. 78-7395.

Softcover reprint of the hardcover 1st edition 1978

2131/3130–543210

Preface

As befits a volume in the Advanced Series in Agricultural Sciences, this book was written with problems of practical agriculture in mind. One of the ways of controlling plant disease is by using resistant cultivars; and from the wide literature of genetics and biochemistry in plant pathology I have emphasized what seems to bear most closely on breeding for disease resistance. This has a double advantage, for it happens all to the good that this emphasis is also an emphasis on primary causes of disease, as distinct from subsequent processes of symptom expression and other secondary effects.

The chapters are entirely modern in outlook. The great revolution in biology this century had its high moments in the elucidation of the DNA double helix in 1953 and the deciphering of the genetic code in 1961. This book, so far as I know, is the first in plant pathology to be conceived within the framework of this new biology. Half the book could not have been written 20 years ago, even if there had then been available all the literature that has since accumulated on the genetics and chemistry of plant disease. The new biology is the cement this book uses to bind the literature together.

Another feature of this book is an emphasis on thermodynamics. Here too I have been fortunate in the timing. During the past decade or two there has been a quickened interest in the thermodynamics of protein polymerization, which is directly relevant to the substance of several chapters. As a result, biochemical plant pathology will, I believe, take a renewed interest in such old-fashioned topics as the effect of temperature on disease.

Proteins are at the center of many of the discussions. This is appropriate, because proteins are the prime link between genes and other molecules, the twin topics of this book. Within the phenotype, proteins are the great stores of mutational change and the governors of mutational effects; and in chapter after chapter proteins emerge as the substance of the discussion, simply as the logical consequence of the experimental evidence.

Variation in the disease resistance of the host and in the pathogenicity of the parasite feature largely, both as the expression of mutation and as the substance of disease control by plant breeding. Chapter 1 starts the discussion; correlated variation in host and pathogen is seen as the downfall of many resistant cultivars, and uncorrelated variation as a source of stability in resistance. Variation threads its way through the chapters that follow, until Chapter 9 knits the threads together in a molecular hypothesis of resistance. The hypothesis is concise, precise, and simple. A great source of confusion about vertical and horizontal resistance has been the lack of a clear chemical explanation of their difference; that explanation is now available.

The final chapter is about biotrophic and necrotrophic processes in parasitism. Nearly a century ago de Bary divided fungi into saprophytes, facultative parasites, and obligate parasites. With various subdivisions and rearrangements, and with reference to necrotrophy and biotrophy, this has been a topic of discussion ever since, without great progress. The fundamental error, I believe, was to try to derive specialized parasitism in the shape of parasitic symbiosis from necrotrophy or from saprophytism. The evidence is, however, that biotrophy and necrotrophy have separate origins, and if this is accepted a fresh understanding of the scope of parasitism is gained. This is not just of academic interest; the choice of the sort of resistance to use in plant breeding is partly determined by biotrophy and necrotrophy.

I am greatly indebted to Dr. D. K. Hallbauer, of the Mining Technology Laboratory of the Chamber of Mines of South Africa Research Organization, for photographs of the precambrian fossil plants *Witwateromyces conidiophorus* and *Thuchomyces lichenoides*.

Pretoria, April 1978 J. E. VANDERPLANK

Contents

Chapter 1 Variation in the Resistance of the Host and in the Pathogenicity of the Parasite

1.1 Introduction 1
1.2 Systems of Variation 1
1.3 Vertical Resistance with Host-Pathogen Specificity 2
1.4 Vertical Resistance Without Host-Pathogen Specificity . . 4
1.5 Epidemiological Consequence of Nonspecific Vertical Resistance 8
1.6 Vertical Resistance Without Hypersensitivity 9
1.7 Correlated Variation in Host and Pathogen 9
1.8 Uncorrelated Variation in Horizontal Resistance 10
1.9 Absence of a Host-Pathogen Differential Interaction in Horizontal Resistance 14
1.10 Uniform Ranking Order in Horizontal Resistance 16
1.11 Tests for Vertical and Horizontal Resistance 19

Chapter 2 The Gene-for-Gene and the Protein-for-Protein Hypotheses

2.1 Introduction 20
2.2 Why Proteins? Why Polymerization? 20
2.3 Flor's Gene-for-Gene Hypothesis 22
2.4 The Living Host Cell Nucleus in Compatible Gene-for-Gene Combinations 23
2.5 The Gene for Susceptibility and Its Different Roles 27
2.6 Comments Arising from the Possibility of Tandem Duplication 30
2.7 The Gene for Avirulence and Its Different Roles 31
2.8 Susceptibility and Resistance. Neutral Mutation 32
2.9 Avirulence and Virulence 33
2.10 Isozymes 34
2.11 The Hydrophobic Effect. Amino Acids with Hydrophobic Side Chains. Their Interrelated Genetic Codes 35
2.12 Amino Acids with Hydrophilic Side Chains. Their Part in Catalysis 38
2.13 Protein Structures 39
2.14 Protein-Protein Recognition 40

Chapter 3 The Protein-for-Protein Hypothesis: Temperature Effects and Other Matters

3.1 Introduction 43
3.2 The Thermodynamic Problem 43
3.3 Inactivation of Protein Monomers 44
3.4 Effect of Temperature on Resistance to Wheat and Oat Stem Rust and Oat Crown Rust 45
3.5 Temperature and Infection of Tomato Plants by Tobacco Mosaic Virus 49
3.6 The Temperature Effect in Some Other Diseases 50
3.7 Some Seemingly Anomolous Examples Considered. 52
3.8 Temperature Reversal 55
3.9 Conditional Responses to Temperature. Dominance of Resistance as a Test 56
3.10 Protein Polymerization and the Solvent 57
3.11 Osmotic Pressure Differences 59
3.12 Analytical Difficulties Arising from Procedures 62
3.13 Isozymes and Electrophoresis 62
3.14 A Hypothesis About Susceptibility and Resistance 63
3.15 Feeding the Pathogen. Function of Genes for Avirulence . . 64
3.16 In Fungus Disease, is Wound Protein the Host Protein Involved? The Function of Genes for Susceptibility 66
3.17 Autoregulation of Gene Expression 67
3.18 A Corollary 68
3.19 Comparison with Antigen-Antibody Systems 69
3.20 Peroxidase Production in Infected Susceptible Plants . . . 69
3.21 The Lag Phase in the Resistance Reaction 69
3.22 Uniformity in Resistance Reaction Types. Secondary Effects 70
3.23 Temperature Limits Again. The Genes *Sr*6 and *Sr*11. Effect of Ethylene 72
3.24 In Virus Disease, Is RNA Replicase the Protein Involved? . . 73
3.25 Summary and Conclusions 75
3.26 Identifying Gene-for-Gene Disease 79

Chapter 4 Common Antigenic Surfaces in Host and Pathogen

4.1 Introduction 83
4.2 Common Antigens in the Flax-Flax Rust System and Some Other Systems 83
4.3 Coexistence of Complementary and Similar Surfaces 84
4.4 The Intrinsic Experimental Difficulty 85
4.5 Alternative Hypotheses About Common Antigens 86
4.6 Common Antigenic Determinants in Other Than Gene-for-Gene Systems 86
4.7 Common Antigens and Male-Female Recognition in Plants 88

Chapter 5 Other Large Molecules in Relation to Gene-for-Gene Disease

5.1 Introduction 90
5.2 DNA 90
5.3 RNA 91
5.4 Glycoproteins 92
5.4.1 The Nature of Glycoproteins 92
5.4.2 Membranes and Membrane Glycoproteins 93
5.4.3 The Carbohydrate Determinants of the AB0 Blood Cell Types 93
5.4.4 One Gene-One Glycosidic Linkage Hypothesis 93
5.4.5 Glycoproteins and the Elicitor Hypothesis 94
5.4.6 An Active Glycoprotein from *Phytophthora megasperma* var. *sojae* 94
5.4.7 A Thermodynamic Objection 95
5.4.8 Elicitors Are Unspecific 95
5.5 Polysaccharides 96
5.6 Lectins 97
5.7 Phytoalexins 98

Chapter 6 Population Genetics of the Pathogen

6.1 Introduction 100
6.2 Directional Selection Towards Virulence 100
6.3 Stabilizing Selection Against Virulence 102
6.4 Nonallelic Virulence Interaction in Stabilizing Selection . . 103
6.5 The Environmental Effect in Nonallelic Interaction Between Virulence on Genes *Sr6* and *Sr9d* 104
6.6 Dissociation of Virulence Through Repulsion 105
6.7 Association of Virulence as an Indirect Consequence of Repulsion 106
6.8 Evidence for Nonallelic Virulence Interactions in Some Other Diseases 108
6.9 A Hypothesis About the Nonallelic Virulence Interaction . . 109
6.10 The Second Gene-for-Gene Hypothesis 111
6.11 The Commonness of Weak Resistance Genes 112
6.12 The Use of Weak Resistance Genes 113
6.13 The Gap in the Use of Vertical Resistance in Plant Breeding . 114
6.14 A Guide to the Use of Vertical Resistance 115
6.15 How to Improve the Performance of Vertical Resistance . . 116
6.16 The Durability of Vertical Resistance 117
6.17 The Effect of Vertical Resistance on Epidemics 118

Chapter 7 Horizontal Resistance to Disease

7.1 Introduction 120
7.2 Horizontal Resistance Not Disguised Vertical Resistance . 121

7.2.1 Hypothesis That Horizontal Resistance Is Buried Vertical Resistance . . . 121
7.2.2 Hypothesis That Horizontal Resistance Is Massed Vertical Resistance . . . 123
7.3 Examples of the Rapid Accumulation of Horizontal Resistance . . . 123
7.4 Polygenic Inheritance . . . 126
7.4.1 Confusion About Polygenes . . . 126
7.4.2 Discontinuous Variation and Qualitative Characters; Continuous Variation and Quantitative Characters . . . 126
7.4.3 Polygenic Variation . . . 126
7.4.4 The Untrue Converse . . . 127
7.5 The Number of Genes Conditioning Horizontal Resistance . . . 128
7.6 The Anonymity of Horizontal Resistance Genes . . . 129
7.7 Specificity in Horizontal Resistance . . . 130
7.8 The Vertifolia Effect . . . 131
7.9 Resistance and Loss of Fitness . . . 132
7.10 Epidemiological Effects of Horizontal Resistance . . . 133
7.10.1 Effect on the Infection Rate . . . 134
7.10.2 Delay of the Start of an Epidemic . . . 134
7.10.3 Contraction of the Scale of Spread of an Epidemic . . . 134

Chapter 8 Selective Pathotoxins in Host-Pathogen Specificity

8.1 Introduction . . . 136
8.2 Some Diseases Caused by Selective Pathotoxins . . . 137
8.2.1 Victoria Blight of Oats . . . 137
8.2.2 Southern Maize Leaf Blight . . . 138
8.2.3 Eye Spot Disease of Sugarcane . . . 138
8.3 Selective Pathotoxins Unconnected with Gene-for-Gene Systems . . . 140
8.4 Evidence for Horizontal Resistance . . . 141
8.5 Horizontal Resistance with a Threshold . . . 142

Chapter 9 A Molecular Hypothesis of Vertical and Horizontal Resistance

9.1 The Hypothesis . . . 143
9.2 Catalysis the Transformer and Eroder of Qualitative Variation . . . 143
9.3 Vertical Resistance: Qualitative Variation . . . 145
9.4 Horizontal Resistance: Quantitative Variation . . . 145
9.5 Vertical Resistance: Reciprocal Host-Pathogen Variation in the Same Class . . . 146
9.6 Horizontal Resistance: Host-Pathogen Variation of Different Kinds . . . 146

Chapter 10 Biotrophy, Necrotrophy, and the Lineage of Symbiosis

10.1 Introduction 147
10.2 Necrotrophy, Biotrophy, and Biotrophy-Necrotrophy Sequences 148
10.3 The Lineage of Symbiosis 150

References 152
Subject Index 165

Chapter 1 Variation in the Resistance of the Host and in the Pathogenicity of the Parasite

1.1 Introduction

One needs to know whether the pathogen can correlate its variation with variation in the host. This need exists irrespective of whether one is concerned with the purely practical aspects of breeding crop plants for resistance to disease, or whether one is probing the genetics and biochemistry of resistance and susceptibility. Accordingly, this chapter, as an introduction to the rest of the book, is given over to a brief discussion of variation in the host and variation in the pathogen, and of whether these variations are correlated. Genes are mentioned in passing, only in so far as they are needed to identify variation, and biochemistry is not mentioned at all.

The system of classifying variation discussed in this chapter is consistent with what is said on genetic and biochemical topics in the other chapters. It determines the order of the chapters, and in Chapter 9 is given an explicit molecular description.

1.2 Systems of Variation

Host plants vary in resistance to disease. If, for example, one examines a large collection of potato cultivars attacked by the blight fungus *Phytophthora infestans*, one finds that some cultivars are severely attacked and some mildly, with many gradations between.

Pathogens vary in their pathogenicity. If, for example, one isolates *P. infestans* repeatedly from a collection of blighted potato cultivars, one finds isolates that differ from one another. The deeper one searches, the more differences one finds: differences in the rate at which mycelium penetrates tubers, differences in the number of sporangia produced per unit area of infected leaf, and so on.

Conveniently we may regard the host variation and pathogen variation as falling within two broad systems. In the one system variation in the pathogen is associated qualitatively with variation in the host. Host varieties have their "own" races of the pathogen, and vice versa. There is an interaction between host varieties and pathogen races. When this occurs, we talk of vertical resistance in the host and virulence in the pathogen.

In the other system, the pathogen varies independently of the differences in the varieties of the host. Host varieties do not have their "own" races of the pathogen. In an analysis of variance, no interaction between host varieties and pathogen races can be demonstrated (provided that appropriate techniques and transformations are used). We then talk of horizontal resistance in the host and aggressiveness in the pathogen.

Basically, the division of resistance into vertical resistance and horizontal resistance follows the same line of thought as that underlying the search for interactions in the analysis of variance. It is, however, usually unnecessary to invoke the procedures of analysis of variance. Ordinarily and historically differential interactions have been demonstrated by the use of differential host varieties. If isolate 1 of the pathogen attacks variety 1 of the host but not variety 2, and if isolate 2 of the pathogen attacks variety 2 of the host but not variety 1, a differential interaction can be accepted as present without further ado. With varying degrees of elaboration this is the general procedure for demonstrating interactions.

Variation in the pathogen correlated with variation in the host implies that if the host is changed, as when a new host variety is brought into cultivation, the pathogen will tend to change too, and so adapt itself to the host. Vertical resistance in the host therefore tends to be "lost" through an associated and matching change in the virulence in the pathogen. This is a practical problem of great importance in plant breeding for resistance. Horizontal resistance, being by definition resistance involved when the variation of the pathogen is not causally associated with variation of the host, is not lost through adaptation by the pathogen.

Vertical resistance and horizontal resistance, as terms, are just definitions. Attempts to prove that horizontal resistance can be lost to new races of the pathogen involve a contradiction in terms. If resistance is lost by parasitic adaptation, i.e., by correlated variation in the pathogen, all other factors remaining constant, it is by definition not horizontal. The real problem is different. It is to decide what resistance is vertical and what horizontal, or, since horizontal resistance is probably always present, what mixture is present of the two forms of resistance.

1.3 Vertical Resistance with Host–Pathogen Specificity

Consider the simplest example, given in Table 1.1. Potato varieties with the gene *R*1 are susceptible to race 1 of *Phytophthora infestans* but resistant to race 2. Varieties with the gene *R*2 are susceptible to race 2 but resistant to race 1. The requirement, that there be relevant variation in both host and pathogen, is met: there are two types of the host and two of the pathogen. The variation of host and pathogen is correlated: a change from an *R*1 type of the host to an *R*2 type is likely to be associated in the course of time with a change from race 1 to race 2 of the pathogen, thus circumventing the resistance.

Table 1.2 is more elaborate, with 16 types of wheat, and three races of *Puccinia graminis tritici*. The races are numbered on the current Canadian system (Green, 1971 b). As with the potato and *Phytophthora infestans* there is a strong differential interaction between wheat and *P. graminis tritici*. Thus, the gene *Sr*6 gives resis-

Table 1.1. The relation between two genotypes of potato and corresponding races of *Phytophthora infestans*

Resistance genes	Race 1	Race 2
R 1	S	R
R 2	R	S

S = susceptible; R = resistant.

Table 1.2. The reaction of 16 genotypes of wheat to three Canadian races of *Puccinia graminis tritici*

Resistance genes	Race		
	C 10	C 33	C 35
Sr 5	S	S	S
Sr 6	R	R	S
Sr 7*a*	R	S	S
Sr 8	R	S	S
Sr 9*a*	S	R	S
Sr 9*b*	S	R	S
Sr 9*d*	S	S	R
Sr 9*e*	S	S	R
Sr 10	S	S	R
Sr 11	S	S	R
Sr 13	S	R	R
Sr 14	S	S	S
Sr 15	S	R	S
Sr 17	S	R	R
Sr 22	R	R	R
Sr *T*2	S	R	R

S = susceptible; R = resistant.

tance to races C10 and C33 but not to C35, while the gene *Sr*9*d* gives resistance to race C35 but not to C10 or C33.

In the potato—*Phytophthora* and wheat—*Puccinia* systems just discussed there is considerable host–pathogen specificity. A host type has a range of pathogenic races to which it is specifically susceptible, and others to which it is specifically resistant. There is an essentially either/or situation. This has made the specification of races relatively easy. If an isolate of *Phytophthora infestans* can infect to potato with the gene *R*1, we assign it to race 1 or to some other race with 1 in its name. So, too, on a more elaborate system one can use 16 or more *Sr* genes to assign an isolate of *Puccinia graminis tritici* to one or other race. The fact that these races are artificial is irrelevant to our narrative. What concerns us here is that the either/or situation of host–pathogen specificity has dominated thinking about vertical resistance, obscuring the fact that vertical resistance exists in other situations.

Even in host–pathogen combinations like a wheat—*Puccinia* combination exceptions to clear specificity occur. With wheat stem rust, for example, it is customary to divide host–pathogen reactions into a series of seven or more classes: "O" (immune), "O;" (nearly immune), "1" (very resistant), "2" (moderately resistant), "3" (moderately susceptible), "4" (very susceptible), and "X" (heterogeneous, intermediate or mesothetic). These seven classes can be extended by adding one or two plus or minus signs to give even finer distinctions. At the two ends of this series, from "O" to "4", resistance and susceptibility are sharply defined. But inevitably as one goes from one end of the series to the other one meets intermediates that cannot be defined as clearly resistant or clearly susceptible. The vertical resistance give by the *Sr* gene is not necessarily tied to sharp host–pathogen specificity. Vertical resistance can occur without host–pathogen specificity.

1.4 Vertical Resistance Without Host–Pathogen Specificity

Consider examples of disease systems in which all races of the pathogen can infect all compatible varieties of the host, but differentially. In such examples one cannot classify a pathogenic race on grounds that it is virulent on a particular resistance genotype. Nevertheless there is vertical resistance, because there is differential interaction and correlated variation.

Xanthomonas malvacearum causes bacterial blight in cotton. In *Gossypium hirsutum* the variety Albar 51 has considerable resistance; and in East Africa several resistant UKA (Ukiriguru Albar) hybrids were produced by backcrossing locally adapted selections with Albar 51. However, the resistance has been matched to some extent by a corresponding change in virulence in the bacterium. Cross (1963, 1964), among others, has investigated the change, both in the greenhouse and in the field.

Table 1.3 summarizes some evidence of Cross (1963). As inoculum he used whole populations of bacteria from stored, air-dried, infected leaf material; and the inoculum was inserted by needle into plants 10–12 days old, just below the cotyledonary node. Lesion length was measured in mm 10, 14, and 18 days after inoculation. Two populations were studied, one from the commercial seed issue, UK 55, which had not been bred for resistance, and the other from UKA 59/513, which had been derived from resistant Albar 51. Of the 18 cotton host varieties studied, four had resistance from Albar 51. The other 14 had not; they were selectively neutral.

Judged by their behavior on the 14 neutral varieties, the two bacterial populations differed little in the lesions they produced. However, on the four varieties selected for resistance the bacterial populations differed greatly. On these varieties the UK 55 population of bacteria caused only small lesion, from 7.2 to 7.8 mm long. The four varieties were resistant to this population, but the UKA 59/513 population caused lesions from 15.5 to 17.3 mm long, which is about as long as those in even the most susceptible of the other 14 varieties. To this population the

Table 1.3. Mean lesion length (mm) for two populations of *Xanthomonas malvacearum* inoculated into the hypocotyls of 18 cotton varieties[a]

Variety	UK 55 population	UKA 59/513 population
A (56)7[b]	7.8	15.5
UKA 67/17/86[c]	7.7	17.3
UKA 121/17/174[c]	7.6	16.2
UKA Ga. 1/17/904[c]	7.2	16.0
Average of 14 others	12.0	12.9

[a] From data of Cross (1963).
[b] A selection from the variety Albar 51.
[c] Selections from crosses with Albar 51.

four varieties had lost their resistance. The bacterial population had become adapted to the host.

Variation in the pathogen was correlated with variation in the host. Vertical resistance occurred; but there was no sharp dividing line between the host varieties or between the bacterial populations. Intermediates occurred, and lesions graded from one to the other. There was no host–pathogen specificity.

Circumstances of survival of X, *malvacearum* favor the adaptation of pathogen to host. The bacteria are carried on and in the seed, and in crop residues which as dust easily blow from field to field. Bacterial populations can cling to a host population to an extent not found in, say, in free-ranging cereal rust fungus which is not seed-borne. The bacteria in a crop are likely to trace their origin back to bacteria on the same host variety the previous season. There is the continuity that facilitates adaptation.

Another example of the adaptation of bacteria to the host, thereby causing vertical resistance to be lost, is that of *Pseudomonas mors-prunorum* on sweet-cherry trees (Crosse, 1975). During 1971 and 1972 in southeast England the level of leaf infection was significantly higher in field trees of the normally more resistant cultivar Roundel than in the normally more susceptible cultivar Napoleon. This reversal of the usual field behavior of the two cultivars was associated with the presence on trees of Roundel of a colony variant of *Ps. mors-prunorum* that was found to differ physiologically and pathologically from typical strains of the organism previously described. Pathogenicity tests with the variant and typical strains revealed highly significant differential interactions between strains and cultivars, with the variant strain showing specialization for Roundel. A similar differential interaction was observed in inoculations of *Prunus avium* seedling clones. One clone, selected repeatedly for high resistance to typical strains, was so badly infected by the variant strain that the trees were irreparably damaged.

The method of survival of *Ps. mors-prunorum* favors continuity from year to year, as it does with *X. malvacearum*. *Ps. mors-prunorum* overwinters in cankers on stems and branches. Bacterial strains can survive and be handed on from season to season on the same tree, thus allowing almost undiluted selection pressure toward the tree's "own" strains of the pathogen.

There are only a few authenticated references in the literature to the breakdown of resistance as a result of variation in bacteria. It may nevertheless well be that breakdown is not uncommon, but that it is often associated with an absence of host–pathogen specificity. Without this specificity, a breakdown might well escape notice unless special tests are made.

It is also possible that breakdowns often occur without host–pathogen specificity in fungus diseases as well. Clifford and Clothier (1974) analyzed data on the variation of *Puccinia hordei* on barley, *Hordeum vulgare*. The cultivar Vada shows some resistance to rust characterized by the slower development of fewer and smaller pustules from a given quantity of inoculum (Clifford, 1972). Clifford and Clothier inoculated leaves of the cultivar Vada with uredospores from five samples of infected leaves of the cultivar Vada, seven of the cultivar Julia and six of the cultivar Sultan. They also inoculated leaves of the cultivars Julia, Sultan, and Proctor. In due course they determined spore production per cm^2 of inoculated leaf surface. They found a highly significant differential interaction between cultivars and the cultures of uredospores.

Consider first the cultivar Vada and the evidence in Table 1.4. The five cultures of *P. hordei* originating from infected leaves of Vada produced an average of 47.7×10^3 spores per cm^2 of inoculated leaf of Vada; but the cultures originating from the other cultivars produced an average of only 31.0×10^3 on Vada. Evidently the cultures from Vada were peculiarly adapted to develop on Vada; and the loss of resistance of Vada when it is infected by its "own" cultures of *P. hordei* is substantial.

When Vada is attacked by its "own" cultures of *P. hordei* the residual resistance, presumably horizontal, is of the same order as that of Julia and Sultan. The relevant figures for spore release in thousands per cm^2 of infected leaf are 47.7 for Vada, 44.9 for Julia, and 48.5 for Sultan, these figures being taken from Tables 1.4, 1.5, and 1.6.

The susceptibility of the cultivar Julia to cultures originating from Julia differs little from its susceptibility to cultures originating from other barley cultivars. The relevant figures are 44.9 and 42.0×10^3 spores per cm^2 (Table 1.5). For the cultivar Sultan the relevant figures are 48.5 and 46.2 (Table 1.6).

Consider the figures for spore production on leaves of the cultivar Julia (Table 1.5) in more detail. The average increase was from 42.0×10^3 per cm^2, when the inoculum originated from other cultivars, to 44.9×10^3 per cm, when the inoculum originated from Julia itself, an increase of 6.9%. Suppose that this increase is real, i.e., ignore for purposes of discussion the matter of experimental error. What then does an increase of 6.9% mean? It means that the basic infection rate R was increased by 6.9%. Suppose the latent period p was 9 days, a figure that would agree with the experimental details recorded by Clifford and Clothier. It is accurate enough to ignore the period of infectiousness. Then by the appropriate equation $r = R\exp[\text{-pr}]$ [Vanderplank, 1963, Eq. (5.9)] applicable to the exponential stage of an epidemic, increasing R by 6.9% would increase the exponential infection rate r from 0.20 to less than 0.21 per day, or from 0.40 to less than 0.41 per day, or from 0.60 to less than 0.61 per day. Such minor increases would have an almost negligible effect on an epidemic. Epidemiologically infection of the cultivar Julia by cultures of *P. hordei* originating from diseased plants of Julia

Table 1.4. Thousands of uredospores produced per cm^2 of leaves of four barley cultivars inoculated with cultures of *Puccinia hordei* isolated from Vada and two other cultivars[a]

Cultivar	Cultures from Vada	Other cultures
Vada	47.7	31.0
Others	43.0	42.8

[a] From data of Clifford and Clothier (1974).

Table 1.5. Thousands of uredospores produced per cm^2 of leaves of four barley cultivars inoculated with cultures of *Puccinia hordei* isolated from Julia and two other cultivars[a]

Cultivar	Cultures from Julia	Other cultures
Julia	44.9	42.0
Others	39.9	41.2

[a] From data of Clifford and Clothier (1974).

Table 1.6. Thousands of uredospores produced per cm^2 of leaves of four barley cultivars inoculated with cultures of *Puccinia hordei* isolated from Sultan and two other cultivars[a]

Cultivar	Cultures from Sultan	Other cultures
Sultan	48.5	46.2
Others[b]	34.9	41.2

[a] From data of Clifford and Clothier (1974).
[b] The disparity between the two figures in this row comes mainly from the inclusion of Vada. Without this cultivar the figures become 39.2 and 42.2, respectively.

would differ only very slightly from infection of Julia by cultures originating from other varieties; and one might conclude that vertical resistance in Julia is small, the bulk of what resistance there is being horizontal. The same would apply to the cultivar Sultan.

There is, however, an alternative. The cultures originating from diseased plants of Julia might not have been of Julia's "own" races. Uredospores might have blown into fields of Julia from fields of Sultan, Vada, Proctor, or any other cultivar, and on the evidence they could easily have started disease in Julia. Internal evidence supports the suggestion that the cultures originating from diseased plants of Julia might well have been heterogeneous. Of the seven cultures originating from Julia, four produced more uredospores on Julia than on the other cultivars averaged, and three produced less. The vertical resistance of Julia (and Sultan) might be higher than the data suggest; about this the evidence leaves us in the dark.

1.5 Epidemiological Consequence of Nonspecific Vertical Resistance

Epidemiologically it is a major difference that, with host–pathogen specificity, a pathogen is necessarily homologous with the cultivar on which it is found—the pathogen necessarily belongs to one of the host's "own" races—whereas, without this specificity, this need not be so. Thus, Vada barley almost certainly has vertical resistance against *Puccinia hordei*, but this is not to say that all cultures originating from a field of Vada belong to "Vada" races. The original source could well have been some other cultivar, the spores being blown into Vada. Possibly because of this Parlevliet (1976c) found substantial stability in the rust resistance of barley cultivars in Europe.

The chance that without host–pathogen specificity an isolate will belong to the "own" race of the cultivar from which it was taken decreases with the mobility of the spores or other inoculum. Wide-ranging uredospores of the cereal rust fungi could cause a cultivar to receive inoculum from many different sources. The chance that a culture will belong to the "own" race of the cultivar from which it was taken also decreases as the degree of multiplication of the pathogen during a season increases. On the experimental evidence, *Phytophthora infestans* can increase a billionfold during a season (Vanderplank, 1960). Inoculum has relatively few sources but very many ultimate destinations. The chance also decreases as the ratio of vertical to horizontal resistance decreases. When vertical resistance is a minor element of total resistance, the importance of a homologous relation between cultivar and pathogen is correspondingly minor.

The relative importance epidemiologically of initial inoculum and infection rate depends on whether there is host–pathogen specificity or not. With host–pathogen specificity, vertical resistance reduces the initial inoculum from which an epidemic starts, but (other things being equal) does not reduce the infection rate at which the epidemic proceeds (Vanderplank, 1968). Without host–pathogen specificity, vertical resistance can reduce the infection rate itself. (See also Sect. 6.17.) In this, vertical resistance without host–pathogen specificity resembles horizontal resistance; but the essential difference between the two sorts of resistance remains: With vertical resistance, irrespective of the presence or absence of host–pathogen specificity, there is a differential interaction between host and pathogen; with horizontal resistance there is not.

It is confusing that many plant pathologists persist in calling vertical resistance specific resistance or race-specific resistance, despite the fact that vertical resistance can occur without host–pathogen specificity. (There is a further objection to the abused term specific resistance being used as a synonym for vertical resistance: horizontal resistance can at times also be specific; see Chap. 7.)

1.6 Vertical Resistance Without Hypersensitivity

A hypersensitive reaction of the host to invasion by the pathogen is not a general feature of vertical resistance when host–pathogen specificity is absent.

Even when there is host–pathogen specificity, hypersensitivity may be absent. Coffey (1976) studied the ultrastructural changes in flax *Linum usitatissimum* after infection by the rust fungus *Melampsori lini*. In flax with the *K* gene for resistance, few of the infected cells became necrotic and collapsed, i.e., few reacted hypersensitively. In most cells the haustorium became encased with a callose-like substance, and up to 9 days after infection most of the cells seemed to be alive. Resistance did not require the death of the infected host cells.

As another example, bean *(Phaseolus vulgaris)* plants of cultivars Canadian Wonder and Red Mexican are resitant of *Pseudomonas phaseolicola* race 1. This resistance is accompanied by a hypersensitive response if relatively high numbers of bacteria are used as inoculum. There is, however, no hypersensitive response following inoculation with low inoculum levels, even though the bacteria multiply to concentrations that would have caused a hypersensitive response had they been used as inocula (Lyon and Wood, 1976). Also, resistance is accompanied by a hypersensitive response if the bacterial cultures are fresh, but cultures more than 8 days old do not cause a hypersensitive reaction, even though the bacteria are viable (Lyon and Wood, 1976).

There is a growing literature of this sort to show that resistance and the hypersensitive response are separate phenomena. The hypersensitive response, the encasement of haustoria or other variations are symptoms of a wrong metabolism of the infected, resistant cells. It will be suggested in the next chapter that with gene-for-gene diseases the wrong metabolism begins with the infected resistant cells' primary coded proteins. A peripheral effect, for that is all a hypersensitive response is, can be expected to vary, as any other symptom can be expected to vary.

One must not, however, jump from one extreme to the other, and assume that because the hypersensitive response is an irregular symptom of resistance, it is not a useful symptom. The hypersensitive response, manifested as a type "1" or "2" reaction of wheat to *Puccinia graminis tritici*, has been used for many years as a trustworthy and entirely adequate indication of vertical resistance. In this example, as in many others, a hypersensitive response implies resistance, even though the processes of resistance and hypersensitivity are governed, genetically and biochemically, as separate sequences. It is the converse, that resistance implies a hypersensitive response, that is false.

1.7 Correlated Variation in Host and Pathogen

The basis of vertical resistance is correlated variation in host and pathogen. When the host changes, then, given the necessary selection pressure, the pathogen will change also.

Gabo and other Australian wheat cultivars carrying the gene *Sr*11 for resistance against *Puccinia graminis tritici* were released in Queensland and northern

New South Wales in 1945. For two years they remained free from stem rust, but in 1948 new virulent races appeared that attacked them. As the acreage of *Sr*11 genotypes increased so did the frequency of isolates of the fungus virulent on *Sr*11. That is, variation in the host population, represented by a change from *sr*11 to *Sr*11 genotypes, was associated with variation in the fungus population, represented by a change from avirulence to virulence on *Sr*11. The selection pressure involved here is the pressure of adaptation of parasite to host.

The wheat cultivar Mengavi, with vertical resistance against *P. graminis tritici* derived from *Triticum timopheevi*, was released in 1960. The correlated variation in the pathogen—the change from avirulence to virulence on Mengavi—was almost immediate.

A more detailed account of adaptation (under the name of directional selection towards virulence) is given for *Phytophthora infestans* in Section 6.2.

With pathogens that disperse widely, as *Puccinia graminis* does, and with host–pathogen specificity, there is a moderately close quantitative relation between the proportion of host plants that carry the gene for resistance and the proportion of isolates from the pathogenic population that carry the corresponding gene for virulence. When a cultivar with a "new" gene for resistance is first introduced into cultivation, the proportion of virulent isolates in the pathogenic population is small. (Otherwise, we assume, the gene would not have been introduced.) Because the proportion of virulent isolates is small, the cultivar is resistant to most of the spores reaching it from other fields, and from the countryside in general. This is the boom period in the cultivar's career. However, if resistance makes the cultivar popular among farmers, its area under cultivation increases and the area under the old cultivars decreases. This increases the proportion of virulent isolates, and resistance is correspondingly lost. This is the bust period in a cultivar's career.

Without host–pathogen specificity, and especially with widely dispersed pathogens, variation in virulence in the pathogen may be expected to lag behind variation in resistance in the host. The boom is not so great, and the bust is not so rapid. However, in the long run, if a cultivar having vertical resistance without host–pathogen specificity comes to dominate the fields, associated variation in the pathogen must inevitably follow, given the necessary selection pressure.

1.8 Uncorrelated Variation in Horizontal Resistance

It is the feature of vertical resistance that, given the necessary selection pressure, variation in the pathogen will tend to follow variation in the host. That was discussed in the previous section. Conversely, it is an essential feature of horizontal resistance that, other factors remaining constant, there is no associated variation of host and pathogen, even though selection pressure exists. Evidence about this is available on a vast scale in the records of blight of potatoes caused by *Phytophthora infestans*. The records come from the Netherlands, and are used as an example.

Table 1.7. The data of introduction in the Netherlands, and the assessments in 1938 and 1968 of resistance to *Phytophthora infestans* in the tubers and foliage of ten cultivars[a]

Cultivar[b]	Date introduced	Resistance assessment[c]			
		Tubers		Foliage	
		1938	1968	1938	1968
Eigenheimer	1893	4	3	4	5
Eersteling	1900	3	3	3	3
Bintje	1910	5	3	3	3
Alpha	1925	8	8	8	7
Bevelander	1925	8	8	8	7
Noordeling	1928	9	8	7	7
Furore	1930	4	6	8	7
Record	1932	8	8.5	8	6
Ultimus	1935	7	7	6	5
Voran	1936	7	6.5	7	7

[a] Assessments recorded in the Nederlandse Rassenlijst voor Landbouwgewassen.
[b] None of these cultivars has an *R* gene.
[c] 10 = very resistant; 3 = very susceptible.

Annually, the authorities in the Netherlands assess the resistance to blight of all registered potato cultivars. The assessment uses a scale in which 10 = highly resistant and 3 = very susceptible. In 1968 there were still ten cultivars that had been assessed continuously for 30 years, and these 30 years had been a period of rapid change both in the cultivars and the races of *P. infestans*. Many cultivars widely planted in 1938 had disappeared from cultivation by 1968; and many cultivars widely planted in 1968 did not exist in 1938. Many of the cultivars in 1968 had *R* genes, which were entirely absent from the cultivars listed in 1938 and therefore absent from the ten cultivars straddling the 30-year period. These *R*-genotypes towards the end of the 30-year period introduced vertical resistance and, in accordance with what was said in the previous section, brought about an associated change in blight races. The changes were large; had there been associated variation of host and pathogen in the ten cultivars, it should have been clearly manifest despite the absence of host–pathogen specificity; but the ten old cultivars went their way over the 30 years without substantial change. Variations in the population of *P. infestans* were not associated with them. The resistance of the ten old cultivars, without *R* genes, can therefore be inferred to be substantially horizontal. An examination of the data has been made (Vanderplank, 1971), but because the evidence is unrivalled in scope, further attention is justified. (The closing date, 1968, has no special significance; it so happened that I spent some months in the Netherlands in 1969, and used the occasion to trace the evidence. No upsetting change has occurred since then.)

Table 1.7 lists the ten cultivars in the order of their coming into cultivation, and gives the official resistance assessments, in 1938 and 1968, in the tubers and the foliage separately. The assessments were necessarily subjective with a change of assessors as time went on, and with a change of resistance background as more resistant cultivars with *R* genes were introduced towards the end of the 30-year period.

To consider tuber resistance first, the average assessment of the youngest five cultivars rose from 7.0 to 7.2 between 1938 and 1968, while the average assessment of the oldest five fell from 5.6 to 5.0. These changes are the wrong way round to support a theory that the cultivars were attacked by their "own" races of blight. The old cultivars in 1938 already had a long history. The average age in 1938 of the oldest three, which are also the three most involved in a changed reassessment, was 37 years; and these three were very popular and widely grown, Bintje in 1938 covering 14000 ha in the Netherlands, with a further considerable area in Belgium and France. Their "own" races, if there were any, should have developed long before 1938. The youngest five cultivars, on the other hand, covered a relatively small area in 1938, although several were to become popular later. On any theory of "own" races, it should have been the younger cultivars that provided the opportunity for new blight adaptations to occur, with a concomitant loss of resistance. The reverse occurred.

If for more comprehensive figures we average the tuber and foliage assessments, we find that the assessment of the oldest five cultivars fell from 5.4 to 5.0 between 1938 and 1968, while with the youngest five the fall was from 7.1 to 6.8. If we use the oldest five to control changes in assessors' standards, there was no evidence for change in the youngest five. If we take the figures at face value, there was a small fall in the assessment of resistance between 1938 and 1968, but it was less than the minimum change, 0.5, that the assessors take notice of. Either there was no change or, if there was, it was too small to be significant in practice.

If instead of listing the ten cultivars in order of age we list them in order of assessed resistance, we again find no great change between 1938 and 1968. Variation in the popularity of the cultivars was not accompanied by detectable variation in the pathogen.

More evidence that variation in the pathogen did not accompany variation in the population of the host to any substantial extent is provided by the cultivar Voran. Introduced in 1936, it covered a relatively very small area in 1937, mostly near fields of the more susceptible cultivar Eigenheimer. However, in the 1940s Voran became very popular in the reclaimed peat districts of the northern Netherlands. In 1950 it covered almost 80% of the total area of about 40000 ha under potatoes in a strip some 80 km long and 20 km wide. Later its popularity started to decline—it was replaced largely by cultivars with *R* genes—and in 1970 was nearly extinct. Here was a great variation in the population of the host: Voran varied from practical insignificance in 1937, to dominance in 1950, and back to insignificance in 1970. Here too, during Voran's dominance, was a massing of a single cultivar in a compact area that must largely have excluded inter-cultivar exchange of inoculum. Yet no detectable relevant variation in the pathogen accompanied this variation in the host's population. Details of the official assessments are given in Table 1.8. There was no fall in assessed resistance as Voran rose to dominance in 1950; nor was there a rise of assessed resistance as Voran sank to obscurity in 1970. Had there been a "Voran" race of *P. infestans* especially adapted to attack Voran, it should have been abundant in 1950 but scarce in 1937 and 1970, with Voran's resistance low in 1950 but high in 1937 and 1970. This did

Table 1.8. The assessment of resistance to *Phytophthora infestans* in the potato cultivar Voran at various dates[a]

Date	Resistance assessment[b]	
	Foliage	Tubers
1937	6	7
1938	7	7
1942	8	7
1950	8	7
1952	8	7
1957	7	6.5
1960	7	6.5
1968	7	6.5
1970	7	6.5

[a] Assessments recorded in the Nederlandse Rassenlijst voor Landbouwgewassen.
[b] 10 = very resistant; 3 = very susceptible.

not happen. The correlated variation of pathogen and host with vertical resistance found no parallel in blight of potatoes without *R* genes in the Netherlands.

It must be said here that the history of blight in Voran as described in the official records and used in the preceding paragraph differs from the history related by Toxopeus (1956). Toxopeus' version is widely quoted in the literature, and the official records ignored. According to Toxopeus, Voran was highly resistant in its early days when, he surmised, it was infected with "Eigenheimer" races of *P. infestans*, but became susceptible from about 1945 when Voran started to be self-supporting in inoculum, i.e., when inoculum reaching Voran fields started to originate from the fungus overwintering in Voran tubers. In comparing the two versions it must be remembered that Toxopeus was reminiscing about events years after they occurred, in order to support his theory of unstable resistance, whereas the official assessors were recording contemporary events year by year without special causes to plead. Details of the two versions are irreconcilable. Thus Toxopeus states that tubers of Voran were not infected in the early years, whereas the official assessment was that they were moderately susceptible (rating 7) from the start. It is unbelievable that such an official assessment could have been published, had tuber infection not been known to occur.

Another potato cultivar commonly quoted e.g., by Toxopeus (1956), along with Voran as having lost its resistance to blight when it became popular is Champion. The evidence is as unconvincing as that about Voran. Champion was released as a commercial variety in 1876, and soon became the most popular variety in Ireland. It had a reputation for considerable resistance to blight. Later it appeared to become more susceptible and lost its popularity. By 1920 it was rated as one of the most blight-susceptible of all varieties, and because of this grown only on a small scale. Toxopeus ascribed the loss of resistance to the rise of Champion's popularity and to the resulting development of a "Champion" race of blight. The question immediately arises, why did Champion not regain its resistance when it lost its popularity? Even more telling against Toxopeus' theory was the work of Davidson (1928). He set out to restore Champion as an Irish cultivar, and to do this he had to free it from potato virus A, with which it had become almost 100% infected. Davidson found that by freeing Champion from virus A he restored its resistance to blight. Champion's story, it seems, is a virus story, not a pathogenic race story.

1.9 Absence of a Host–Pathogen Differential Interaction in Horizontal Resistance

From the definition, one would expect a host–parasite differential interaction to be absent when only horizontal resistance is involved. Consider this example: Paxman (1963) set out to determine whether races of *Phytophthora infestans* would become specially adapted to a potato cultivar, if they were grown continuously on it. He used cultivars without *R* genes. He obtained an isolate 30*RS* from a naturally infected tuber of the cultivar Red Skin, another isolate 31*KP* from a naturally infected tuber of the cultivar Kerr's Pink, and an isolate 32*KE* from a naturally infected tuber of the cultivar King Edward. These he subcultured, each on its original cultivar, i.e., he subcultured isolate 30*RS* on tubers of Red Skin, 31*KP* on tubers of Kerr's Pink, and 32*KE* on tubers of King Edward. He started his test after 90 cycles of subculturing. That is, the isolate 30*RS* had been on the cultivar Red Skin for 90 cycles of subculturing plus the unknown number of cycles that the fungus had been on Red Skin before it was isolated. He used as his criterion the rate of spread of each of the three isolates in tubers not only of the cultivar of origin, e.g., isolate 30*RS* in Red Skin, but also in tubers of the other two cultivars, e.g., isolate 30*RS* in tubers of Kerr's Pink and King Edward as well. He also used a fourth isolate of unspecified origin which he cultured on tubers of the cultivar Majestic. His results show a highly significant difference between isolates. That is, the isolates differed significantly in their ability to spread through tuber tissue. There was also a highly significant difference between cultivars. That is, cultivars differed significantly in the resistance they offered to mycelium spreading through tuber tissue; but—and this is what concerns us most—there was no evidence for a cultivar x isolate interaction. The pathogenicity of the isolates was uniformly spread over the host varieties; and the resistance of the host varieties was uniformly spread against the isolates.

Paxman showed that, under the conditions he used, he had been unable to adapt *P. infestans* to a potato cultivar. He did not, however, prove that races of *P. infestans* are not specially adapted to particular cultivars. This goes beyond the scope of his data. To prove the absence of special adaptation he would have had a show that the isolate 30*RS* was in fact the cultivar Red Skin's "cwn" race. There is no proof of this. Inoculum might have blown into the field of Red Skin during the season from a field of some other cultivar; indeed, it is quite likely that it did, because Red Skin is a maincrop variety and annual epidemics commonly start in the early-maturing cultivars. So, too, it is open to doubt whether the isolates 31*KP* and 32*KE* had any real connection with the cultivars Kerr's Pink and King

Edward. Their presence in these varieties could have been a matter of chance and the vagary of wind direction during sporulation. What Paxman did show, and what he set out to show, was that he could not specially adapt *P. infestans* to a variety during the period of his culturing.

Results different from those of Paxman have been published by Jeffrey et al. (1962), Jinks and Grindle (1963), and Caten (1974), all working with material from plots in an experimental field in Birmingham. Jeffrey et al. collected isolates of *P. infestans* from infected leaves of nine varieties in these plots. The varieties were eight potato cultivars and an unidentified species of *Solanum*, all without an *R* gene. Each isolate was grown in tubers of the varieties Arran Consul, Majestic, and the variety from which it was originally isolated. They measured the penetration of mycelium at 17 to 19° C, and the metric chosen for analysis was the growth in mm over 10 consecutive days. They found a highly significant interaction between varieties and isolates. The growth of all isolates on their "own" varieties was faster than on Arran Consul or Majestic, except when the isolates came from either of these two varieties. A further sample of 14 isolates made in the same year confirmed their superiority on their "own" varieties, as did two samples of nine isolates in a later year.

The difficulty in accepting these results is both experimental and statistical. On the experimental side it is difficult to accept the basic assumption that each plot in the experimental field at Birmingham was infected with the variety's "own" race of *P. infestans*: the Majestic plot with the "Majestic" race, and so on. How did each race come to be in its own variety? Transmission of *P. infestans* through tubers used as seed is a rare event. The first detailed observations were those of van der Zaag (1956) who found an average of one seed-tuber transmission of *P. infestans* per km^2 of potato fields of the very susceptible variety Duke of York, and even less transmission in less susceptible varieties. Since then other surveys, both on the ground and by aerial photography, have confirmed that transmission through infected seed tubers is a rare event. In Maine, Bonde and Schultz (1943, 1944) concluded that transmission through seed tubers was so rare that inoculum for the initial blight outbreaks in fields came from the piles of culled potatoes discarded after the winter seed stores had been opened. All in all, the evidence is convincingly against infection coming via seed into each variety plot, even once. It is even more strongly against infection coming into each plot often enough to saturate it with blight, and thus exclude inter-plot movement of *P. infestans* in the foliage. On this last point, those with experience of blight in experimental plots of a variety collection will have observed initial foci of blight enlarging and engulfing all varieties (without *R* genes) in their path, all varieties engulfed in a single focus being presumably infected with *P. infestans* of a single common origin. It is unnecessary to elaborate further objections to the basic assumption that each variety somehow became infected with its "own" race. The onus was clearly on Jeffrey et al. to explain how each variety came to have its own race, against all available epidemiological evidence. They did not do so, nor did they give details about the size and shape of the plots, which are relevant to their assumption.

Johnson and Taylor (1976) sought to counter this objection by arguing that in a mixture of races in a cultivar plot the cultivar's "own" race would in the course of time outstrip its competitors and take over. The argument does not touch the objection. Irrespective of whether there are other races or not in a cultivar plot, the question remains: how did the cultivar's "own" race get into the plot in the first place? Where, for example, did the unidentified species of *Solanum* get its "own" race from, at any stage in the course of the season?

A possible way out would be to assume that *P. infestans* quickly adopts another variety as its "own", by becoming trained to it. This assumption is ruled out by Jeffrey et al. (1962), Jinks and Grindle (1963) and Caten (1974) who were unable to train isolates of *P. infestans* to adopt other potato varieties as their "own". On this matter their results agree with those of Paxman. It is perhaps more than a concidence that the only matter about which the experimenters agree is the only matter about which it is irrelevant: whether the isolates from varieties were the varieties' "own" or not.

There is a statistical difficulty as well. If of nine potato varieties, chosen randomly from many, each had its own clearly distinguishable race of *P. infestans*, then, within conventional confidence limits, out of every million potato seedlings of random parentage at least 700000 would each have its own distinguishable race. (Some potato breeding organizations raise more than a million seedlings a year; and the number of seedlings that would be needed to leave a residue of nine cultivars must be very large.) It does not matter that few, if any, of the 700000 seedlings would survive the selection processes to become cultivars; the potential of *P. infestans* to match each of them would in any case have to be there. The numbers involved are impressive; but the ratio 10:7 is even more impressive. For every ten potato variants, the variation being related to any character, agronomic or otherwise, there would have to be seven variants of *P. infestans*, the variation relating to pathogenicity alone. Most potato characters, e.g., the shape of the tuber, are not known to be related to susceptibility to blight; and even if one allows for the fact that genes commonly have multiple effects, the ratio 10:7 needs to be explained before the evidence about "own" races can carry conviction.

To avoid misunderstanding let it be said that it has not been the intention of this chapter to try to prove *in principle* that resistance to blight in potatoes without *R* genes is purely horizontal. On that matter one must keep an open mind and judge by the evidence. There is no a priori reason why vertical resistance should not be found in cultivars without *R* genes, and it would make little difference to the theory of resistance if it were found. The intention is different. It is commonly assumed in the literature that vertical resistance must always occur, and that the pathogen must inevitably adapt itself qualitatively to the host and overcome its resistance. This is the defeatist principle that is being discussed and disputed; and we let the evidence, in this and the previous section as well as in the next, speak for itself.

1.10 Uniform Ranking Order in Horizontal Resistance

In Section 1.8 horizontal resistance was considered as variation in the population of the host uncorrelated with a corresponding variation in the population of the pathogen, despite selection pressure towards association. In Section 1.9 horizontal resistance was considered as the absence of a differential interaction between varieties of the host and relevant isolates of the pathogen. In the present section horizontal resistance is considered as a uniform ranking order of races of the pathogen in their aggressiveness on different host varieties, or, alternatively, as a uniform ranking order of host varieties in their resistance to different races of the pathogen. In all three sections the discussions are equivalent. We are dealing, not with three definitions of horizontal resistance, but with three different manifestations of horizontal resistance as a single uniform concept.

Johnson and Law (1975) studied three genotypes of wheat and five races of *Puccinia striiformis*, the cause of stripe rust. This section is based on their experiments.

The cultivar Hybride de Bersée (referred to here as euploid Bersée) of hexaploid wheat (*Triticum aestivum*) has the reputation of being moderately resistant to stripe rust. It has 21 normal pairs of chromosomes. The chromosomes carry a reciprocal translocation with respect to chromosomes 5B and 7B of the cultivar Chinese Spring. The translocation is such that one chromosome of euploid Bersée corresponds with the long arms of chromosomes 5B and 7B in Chinese Spring and is therefore designated 5BL-7BL, while another in euploid Bersée corresponds with the short arms of these chromosomes and is designated 5BS-7BS. The second host genotype for this discussion is Hybride de Bersée monosomic 5BS-7BS (referred to here as mono 5BS-7BS). It has 20 pairs of chromosomes plus a single short chromosome which corresponds with the short arms of chromosomes 5B and 7B in Chinese Spring. The third wheat genotype is Hybride de Bersée nullisonic 5BS-7BS (referred to here as nulli 5BS-7BS). It has 20 pairs of chromosomes, lacking the pair 5BS-7BS. The three genotypes are all vertically resistant to race 37E132, but the genes for this resistance are on chromosomes other than 5BS-7BS, and do not enter the story. To the five races with which we are concerned, races 40E8, 41E136, 104E9, 104E137, and 108E9, the three wheat genotypes have no vertical resistance; and the reaction types are "3" to "4".

In one experiment, using seedlings, Johnson and Law estimated spore production, determined as mg of uredospores per 100 cm^2 of leaf, on euploid Bersée and nulli 5BS-7BS, using cultures of three races. There was a highly significant difference between the two host types, more than twice as many spores being produced on nulli 5BS-7BS as on euploid Bersée. There was also a great difference between cultures of the pathogen, cultures of race 40E8 producing three times as many spores as those of race 41E136. There was, however, no significant interaction between wheat types and race cultures. There was therefore no evidence for resistance other than horizontal resistance.

In another experiment, using adult plants, Johnson and Law estimated the percent leaf area infected in the field. Table 1.9 gives mean estimates for the three different dates of assessment, for three host types and five cultures of the pathogen.

These data can be used to test for a differential interaction between types of the host and cultures of the pathogen, by ranking the host types in order of decreasing resistance to the various cultures (Table 1.10) or by ranking the cultures in order of increasing aggressiveness on the various host types (Table 1.11).

Table 1.9. Mean percent leaf area infected in field tests of three wheat host types and cultures of five races of *Puccinia striiformis*[a]

Host	Race				
	40E8	41E136	104E9	104E137	108E9
Euploid Bersée	3	<1	< 1	<1	< 1
Mono 5BS–7BS	15.8	1.2	1.5	1.5	3.7
Nulli 5BS–7BS	33.3	2.3	13.3	8.2	23.3

[a] Means of two plots and three dates, June 27, July 7, and July 20, from data of Johnson and Law (1975).

Table 1.10. Three wheat host types ranked in order of decreasing resistance to cultures of five races of *Puccinia striiformis*[a]

Host	Race				
	40E8	41E136	104E9	104E137	108E9
Euploid Bersée	1	1	1	1	1
Mono 5BS–7BS	2	2	2	2	2
Nulli 5BS–7BS	3	3	3	3	3

[a] Data of Table 1.9, reworked.

Table 1.11. Cultures of five races of *Puccinia striiformis* ranked in order of increasing aggressiveness to three wheat host types[a]

Race	Host		
	Euploid Bersée	Mono 5BS–7BS	Nulli 5BS–7BS
41E136	1.5	1	1
104E137	1.5	2.5	2
104E9	3.5	2.5	3
108E9	3.5	4	4
40E8	5	5	5

[a] Mean values for June 27 and July 7 only. These give more information than the means shown in Table 1.9. 1 = least aggressive; 5 = most aggressive. 1.5 means a tie for the first and second places.

There is no evidence in these tables of any significant departure from uniformity; resistance was substantially horizontal.

There are biometric methods available for measuring the degree of uniformity between two or more rankings, i.e., for measuring the degree of rank correlation, but in Tables 1.10 and 1.11 the correlation is so obvious as to need no sophisticated statistical probe. Ranking methods are distinct in that they involve counting, not measuring; and this brings with it distinct advantages and disadvantages. This matter is taken up briefly in the next section.

There is a clear case that resistance both in seedlings and adult plants of the cultivar Hybride de Bersée is substantially horizontal, and that the factors governing this resistance are largely on chromosome 5BS-7BS. Johnson and Law dispute that the resistance is horizontal, "because cultivars have greater resistance to some races of *P. striiformis* than to others." There is a misunderstanding here that needs to be cleared up. Consider this example: the cultivar Hybride de Bersée is more resistant to race 41E136 of *P. striiformis* than to race 40E8. This, Johnson and Law would argue, rules out horizontal resistance. Observe, first, that strictly considered, the example compares aggressiveness, an attribute of the pathogen, and not resistance, an attribute of the host, because the example concerns only one host type but two pathogenic races. Observe, secondly, that from this example it would be impossible to decide what sort of resistance is involved—whether it is horizontal or vertical or any other sort—because there is no variation in the host, there being only one host type named in the example. To determine whether resistance is horizontal or vertical there must be relevant variation in *both* the host and the pathogen (see also Sec. 3.26) and if cultivars have greater resistance to some races of *P. striiformis* than to others, this is a necessary ingredient of horizontal resistance, and not, as Johnson and Law believe, an objection to it.

1.11 Tests for Vertical and Horizontal Resistance

Johnson and Taylor (1976, their Table 1) have neatly demonstrated how in the analysis of variance a fictitious interaction, pathogenic isolates x host varieties, may arise for purely arithmetical reasons. The difficulty can be overcome by appropriate transformations of the variate (in Johnson and Taylor's example, by the use of logarithms). Alternatively it can be overcome by following the practice in this book and its predecessor (Vanderplank, 1975): Confine the use of the analysis of variance to evidence for the *absence* of vertical resistance, and use other methods to demonstrate its presence.

Elegant tests of significance exist for ordinal numbers 1, 2, 3... which denote order or rank. Ranking methods lose some information, in that they discard information about how close the various ranked members may be within the scale. In this they are inferior to the familiar statistical methods that use cardinal numbers. However, ranking methods have the unique advantage that ranking stays independent of whether the scale of measurement is stretched or compressed. For example, plant pathologists commonly assess the severity of disease on an arbitrary scale, with, say, 0 = very susceptible, 10 = very resistant. If, for example, the inherent biological stretch between assessments 5 and 6 differs from that between assessments 8 and 9, ranking remains valid, whereas conventional methods involving cardinal mean values, etc., become suspect. So, too, ranking methods can be applied to estimates of percentage disease, without transformations that might or might not be apt and adequate. A study of techniques of ranking analysis in plant pathology could be a profitable topic of research, both generally and in relation to vertical and horizontal resistance.

Chapter 2 The Gene-for-Gene and the Protein-for-Protein Hypotheses

2.1 Introduction

In diseases in which host and pathogen are involved gene for gene, susceptibility involves the copolymerization of protein from the host with protein from the pathogen. In gene-for-gene diseases host and pathogen recognize each other by their proteins. This hypothesis was stated in a brief essay (Vanderplank, 1976). Here we fill in details.

This chapter and the next deal with the hypothesis. In this chapter the general background is given; in the next, a long chapter, a mass of familiar experimental data are discussed. These data have been accumulated in the literature for more than half a century, but largely neglected because their relevance to our topic was missed.

It is necessary in this chapter to introduce topics such as the thermodynamics of protein polymerization, in which some plant pathologists might not be well versed. For this reason it could perhaps be useful in a first reading to concentrate on familiar topics, skip the essentially chemical sections, and pass on to Chapter 3.

In the title of this chapter and generally in the text, we use the word protein more often than enzyme. Enzymes are involved, but mainly through their properties as proteins, and only to a lesser extent through their special property as catalysts. The more embracing word protein is therefore apter in most, but not all, contexts. We interchange the words as required.

2.2 Why Proteins? Why Polymerization?

We are concerned with the chemical storage of variability. With wheat stem rust, for example, if we allow for 20 genes for resistance in the host and 20 for virulence in the pathogen, there are potentially at least 2^{20}, roughly a million, pathogenic races. Each race has its own individual constitution of genetic variation, which reflects genetic information stored in the genes of both host and pathogen.

Only three known classes of substance are capable of storing information of variation on this scale. They are, first, the nucleic acids themselves, second, proteins, and, third, complex carbohydrates, including glycoproteins in which the carbohydrate moiety carries the relevant variation. Each of these classes must be studied impartially. In this and the next chapter we study proteins, in Chapter 5 nucleic acids and complex carbohydrates.

Proteins can conserve most of the information about mutation. Axiomatically, all the information about mutation is stored in the nucleic acids, and some is lost with the sequence of synthesis; but the loss between nucleic acid and protein is relatively small. Of information about random mutations involving base substitutions, approximately three fourths passes from the nucleic acids to the coded proteins. To be more precise, of every 24 random mutations involving base substitutions, 17 on an average are missense mutations (Whitfield et al., 1966) that pass on to the proteins in the form of altered amino acids. (Of the rest, 6 of the 24 are samesense mutations which are not passed on to the proteins because they leave the amino acids unaltered, and 1 is a nonsense mutation that cannot specify an amino acid at all.)

In synthesis beyond the proteins the loss of information is severe. The products of proteins, through enzyme catalysis, reflect little of the proteins' variation. An enzyme is a globular protein built up, commonly, of several hundred amino acid residues of which relatively few are bound to the substrate or directly involved in catalysis. The rest of the amino acid residues, if they vary, vary as isozymes, which by definition are genetically determined variations of an enzyme that do not change the product of catalysis, although they may change the efficiency of catalysis. The abundance of isozymes therefore measures the abundance of qualitative variation that never gets beyond the enzymes. This matter is elaborated in Chapter 9.

Proteins are thus strategically placed in the center of the sequence of synthesis. They receive most of the information on variations from the nucleic acids, and chemically are vastly more versatile and reactive than the nucleic acids in making use of the information. On the other side, they filter out most of the information on qualitative variation, and make their products relatively poor storehouses of genetic variation.

Why polymerization? The answer is, to bring the protein-stored information from host and pathogen together without loss through catalysis. When we say that there are potentially a million races of wheat stem rust we are saying that there are potentially a million differential interactions between wheat and *Puccinia graminis tritici;* and when we say that there are potentially a million differential interactions, we are saying that chemical contact between wheat and *P. graminis tritici* can vary in potentially a million qualitatively different ways. The contact is real. There is no action at a distance in chemistry, i.e., no action outside the range of chemical bonds. In terms of our topic, host protein and pathogen protein must make physical contact with each other, without their products being intermediaries, i.e., without the loss of variation that catalysis involves. Direct contact without catalysis implies polymerization—there is no other known form—and enough is known of mutual recognition of protein–protein surfaces to leave little doubt that polymerization could explain the occurrence of pathogenic races by the million.

It is customary to use the words polymer and polymerization when the monomers or sub-units are similar, and the words copolymer and copolymerization when they are not. Chemically, polymerization and copolymerization are similar, and when discussing the chemical process we tend to use the shorter word; but when the point of the discussion is that the copolymer is derived partly from the host and partly from the pathogen, which is what the protein-for-protein hypothesis is about, the longer word is more apt in that it is noncommital about whether the sub-units are similar or not.

We apply the protein-for-protein hypothesis to diseases to which the gene-for-gene hypothesis has been applied. Further applications must be left to the future. The two hypotheses differ somewhat in emphasis. The gene-for-gene hypothesis centers around genes for resistance in the host plants. The protein-for-protein hypothesis is concerned primarily, but not exclusively, with susceptible host plants, i.e., with compatible host–pathogen combinations. Examples of gene-for-gene diseases were given in the previous chapter, in Tables 1.1 and 1.2.

2.3 Flor's Gene-for-Gene Hypothesis

Flor (1942) was the first to study the genetics of both members of a host–pathogen system, the host being flax, *Linum usitatissimum*, and the pathogen the flax rust fungus, *Melampsora lini*. In flax varieties possessing one gene for resistance to the avirulent parent race, pathogenicity was conditioned by one gene in the fungus. In flax varieties possessing two, three, or four genes for resistance to the avirulent parent race, pathogenicity was conditioned by two, three, or four genes in the fungus. The hypothesis that for each resistance gene in the host there is a matching and reciprocal gene for pathogenicity in the fungus is the simplest that fits the facts.

The concept of a gene-for-gene relation has been applied, with varying degrees of proof, to other host–pathogen combinations, the pathogens including viruses, bacteria, fungi, nematodes, insects, and a flowering plant (*Orobanche*). Lists, with relevant literature citations, have been compiled by Flor (1971), Person and Sidhu (1971), and Day (1974). Table 2.1 shows some of the combinations that have been

Table 2.1. Shortened list of host-parasite systems for which gene-for-gene relations have been suggested or proved

Avena	*Puccinia graminis avenae*
Gossypium	*Xanthomonas malvacearum*
Hordeum	*Ustilago hordei*
Leguminoseae	*Rhizobium*
Linum	*Melampsora lini*
Lycopersicon	*Cladosporium fulvum*
Lycopersicon	Tobacco mosaic virus
Malus	*Venturia inaequalis*
Solanum	*Heterodera rostochiensis* (golden numatode)
Solanum	*Phytophthora infestans*
Solanum	*Synchytrium endobioticum*
Triticum	*Erysiphe graminis tritici*
Triticum	*Puccinia graminis tritici*
Triticum	*Puccinia recondita*
Triticum	*Puccinia striiformis*
Triticum	*Tilletia caries, T. contraversa*
Zea	*Puccinia sorghi*

demonstrated or suggested. It has been shortened to include only diseases mentioned elsewhere in this book.

It is inherent in the hypothesis that a gene-for-gene combination can be detected only when there is a relevant resistance gene in the host. There must be a large number of buried gene-for-gene relations which have escaped notice for lack of a resistance gene; and even the most detailed published lists of combinations probably give a highly incomplete picture of the scope of gene-for-gene relations. Thus, *Solanum tuberosum sensu stricto* has no gene of resistance to *Phytophthora infestans* known to qualify for the gene-for-gene hypothesis. It needed the gene *R*1 from *S. demissum* to be transferred to *S. tuberosum* before it could be suggested that *S. tuberosum–P. infestans* was a gene-for-gene combination. Other resistance genes, *R*2, *R*3, ..., have also been transferred to the potato; it now seems possible, though it has not been proved, that there are eleven or more loci in *S. tuberosum* which could qualify for gene-for-gene relations with *P. infestans*. Plant breeders had to use resistance genes from Central American *Solanum* spp. before blight in *S. tuberosum*, a South American species in origin, joined the list of suspected gene-for-gene diseases.

So too with wheat and *Puccinia graminis tritici* the demonstration of a gene-for-gene relation depended on the discovery of resistance genes. In the days when the common wheat varieties were universally susceptible, there was no reason to suspect a gene-for-gene relation. It was only through the incorporation of stem rust resistance genes—*Sr* genes—one after another that the extent of the gene-for-gene relation could be envisaged. In general, it is among those diseases which plant breeders have set out to control through incorporating resistance genes that experimental evidence for the gene-for-gene hypothesis has accumulated.

Nevertheless, despite the known gene-for-gene diseases being probably only a small sample of the total, we shall use that sample as being probably fairly representative of gene-for-gene associations in general.

2.4 The Living Host Cell Nucleus in Compatible Gene-for-Gene Combinations

It is essential for the protein-for-protein hypothesis that in compatible (susceptible) host–pathogen combinations the host cell nucleus should survive infection of the cell, at least for a short while. The nucleus (together, perhaps, with cytoplasmic sources of DNA) specifies the protein that, according to the hypothesis, copolymerizes, i.e., enters into a protein-for-protein association. Its survival, at least until an association is established, is an integral part of the hypothesis; and its survival is a demonstrable fact in known gene-for-gene diseases.

The universality of the survival of the compatible host nucleus in a gene-for-gene combination can be tested by using Day's (1974) list as the sample. In every one of the 27 combinations listed by him, the host nucleus survives to a greater or lesser extent. We need not here analyze every combination, but shall consider a few examples to show the range of adaptations.

Table 2.2. Effect of infection by *Puccinia recondita* on the survival and size of the nuclei of cells of Little Club wheat[a]

Age of infection (days)	Nuclei survived %	Size of living nuclei (μ)	
		Infected cells[b]	Normal cells
5	98	8.8 × 7.0	8.2 × 7.1
9	93	10.1 × 6.3	8.6 × 6.5
12	98	10.6 × 6.3	8.5 × 6.2
16	91	9.1 × 7.4	8.7 × 7.4

[a] Compiled from data of Allen (1926).
[b] Living cells at the center of the lesion.

Consider, first, leaf rust of wheat caused by *Puccinia recondita.* A gene-for-gene relation was demonstrated by Samborski and Dyck (1968, 1976), and Bartos et al. (1969). Allen (1926) studied the infection process in the fully susceptible wheat variety Little Club. The germinating uredospores infect through the stomata, and the branching hyphae quickly establish haustoria in the host cells. In no case observed, up to the sixth day at least, does the host cell suffer visibly from the haustorial invasion. It is not plasmolyzed, collapsed, or even impoverished. The only disturbance noted is in the position of the host nucleus, which seems to be powerfully attracted by the haustorium and may move towards it. When two haustoria lie a short distance apart, the host nucleus is often midway between the two. When nearer together, the nucleus may be drawn out into a dumbbell-shaped body having contact with both haustoria. In a few instances Allen observed the host nucleus wrapped around the young haustorium, partly enclosing it. This contact between host nucleus and parasite haustorium does no visible harm to either body, and both continue to live and function. Spore production by the fungus begins early, and the first spores are liberated on the 7th, 8th, or 9th day after infection. Infected tissues live on, and even 16 days after infection, i.e., 7–9 days after spore dispersal begins, not more than a few of the infected cells are dead. Chester (1946) records spore production persisting for about 2 weeks.

The host nucleus in wheat leaf rust infections enlarges, but only slightly, and survives well (Table 2.2). The whole picture is one of substantial compatibility between host and parasite.

There are other examples of great compatibility between cereal hosts and rust fungi. Durrell and Parker (1920) observed lesions of *Puccinia coronata* on oat plants producing spores continuously for a month; and Cammack (1961) observed lesions of *P. polysora* on maize producing spores copiously for 18 to 20 days.

There is a large literature of evidence for a gene-for-gene relation between wheat and *Puccinia graminis tritici* (Luig and Watson, 1961; Loegering and Powers, 1962; Green, 1964, 1966; Williams et al., 1966; Kao and Knott, 1969). In this combination, too, the invaded protoplasts of susceptible host plants mostly survive for many days after sporulation starts; and although the congeniality between host and parasite is not as great with *P. graminis tritici* as with *P. recondita*, there is no doubt that wheat stem rust conforms with the general rule about the survival of the nucleus in infected cells of compatible host plants belonging to gene-for-gene combinations. Differences between *P. graminis tritici* and *P. recondita* are in detail. There seems to be a greater enlargement of the nucleus of host cells invaded by *P. graminis tritici;* and by the 15th day after invasion of the host protoplast, many nuclei have degenerated in stem rust, but few in leaf rust.

The findings of Rohringer and Heitefuss (1961) and Whitney et al. (1962) explain much. The initial swelling of the host nucleus after the invasion of the protoplast by *P. graminis tritici* is accompanied by a swelling of the nucleolus, and an increase of activity of the RNA but not the DNA. We note, first, that the nucleolus is a center of RNA and protein synthesis; and increases in nucleolar size and RNA content are symptomatic of cells actively engaged in protein synthesis. A parasitized host cell that can be stimulated to produce more protein is being stimulated to produce more food for the parasite, and to produce it in a way most relevant to a protein-for-protein system. Secondly, we note the absence of an increase in the DNA content of the host nuclei of parasitized cells. The cells do not divide, and there is no evidence that the DNA itself, as distinct from its coded products, plays a direct chemical part in the parasitism of wheat stem rust.

The host cell nucleus survives until after sporulation in wheat leaf rust and stem rust, and this is typical of many other fungus diseases which have been classed as gene-for-gene diseases: other rust diseases, powdery mildews caused by *Erysiphe* spp., apple scab caused by *Venturia inaequalis*, and tomato leaf mold caused by *Cladosporium fulvum*. In some other fungus diseases classed as gene-for-gene diseases, the nucleus survives invasion of the protoplast, but only until the fungus begins to sporulate, or just before.

Phytophthora infestans in *Solanum tuberosum* has been classed as a gene-for-gene disease by Black et al. (1953) and Toxopeus (1956). In young infections of compatible host varieties the nucleus survives for a while after the host cell has been invaded by haustoria and transcellular hyphae (Hohl and Suter, 1976); but it does not generally long survive sporulation. The effect of sporulation is perhaps best seen macroscopically. In a developing lesion on a leaf there is an outer zone bordering on the surrounding healthy tissue. In this outer zone the protoplasts and nuclei are alive. Within this zone is a sporing zone about 7 to 10 mm wide. Sporulation in this zone continues for about a day, after which the zone moves outward into previously nonsporing infected tissue. Inside the sporing zone is an ever widening zone of necrotic, sterile tissue, with nucleus and protoplast dead. The macroscopic effects have been studied in detail by Lapwood (1961a, b, c).

In the smuts and bunts sporulation marks the end of the living host cell. *Ustilago hordei* in *Hordeum* has been shown by Sidhu and Person (1972) to be a gene-for-gene disease. Barley seedlings are infected systemically; and systemic infection continues without apparent damage to the host nucleus until the pathogen destroys the seed head to produce teliospores.

Synchytrium endobioticum in *Solanum* has been suggested by Howard (1968) to be a gene-for-gene disease. Infection is intracellular, and parasite and host nucleus live side by side for a time; but the beginnings of sporulation see the end of the host nucleus. When reproduction is asexual, the host nucleus does not survive the development of the prosorus from which sporangia and zoospores are formed.

When reproduction is sexual, the nucleus of the invaded host cell stays active long enough for one or two cell divisions to take place, but not much longer.

Bacteria infecting parenchyma tissue of compatible host plants remain intercellular, and the host cell's nucleus remains alive until the cells collapse. *Xanthomonas malvacearum* in cotton (*Gossypium* spp.) has been identified as a gene-for-gene disease by Brinkerhoff (1970), and the infection process has been studied by Cason et al. (1977). They inoculated the cotyledons of a susceptible cotton variety with bacteria. Microscopic examination of sections of the cotyledons six days after inoculation showed bacteria densely distributed through the intercellular spaces of the palisade and spongy parenchyma cells, with no apparent reaction of the host nucleus to pathogenesis. Thereafter the bacteria dissolved the cells. The condition is met: the host cell nucleus in infected tissue remains alive, at least for a while after infection.

The reference in the preceding paragraph was to infected tissue, not infected cells. Bacteria have no haustoria or other apparatus to penetrate living host cells of the parenchyma. Until the host cells collapse, the bacteria remain in the intercellular spaces. Nevertheless, as with fungi having haustoria, there is clearly an interchange of substances between host and pathogen. From the susceptible host, through its cell walls, comes enough nutriment to support a dense mass of bacteria in the intercellular spaces. From the bacteria, in resistant (incompatible) host varieties, comes enough substance to provoke a hypersensitive reaction within the host cells 4 h after inoculation. In relation to the protein-for-protein hypothesis there is nothing in principle to separate bacteria and fungi.

X. malvacearum, as Cason et al. showed, confines itself to the parenchyma, where the cells have nuclei, and does not enter the xylem, where the cells lack them. This tallies with the hypothesis. It also explains the original name of the cotton disease, angular leaf spot, the area of the primary infections being limited by the small leaf veins.

Nutman (1969) suggested that the recognition of Leguminosae by *Rhizobium* is on a gene-for-gene basis. Rhizobia enter the root via the root hairs, and then make their way through the middle lamellae of the root cortical cells. The adjacent cells show many cytoplasmic changes (Tu, 1976), even though the cytoplasm at this stage is not invaded. These changes include an enormous increase in the number of ribosomes, which suggests increased RNA and protein activity. Later, the rhizobia invade the cytoplasm of the cortical host cells by endocytosis. The nucleus remains active, and the invaded host cells divide, as do the bacteroids along with them.

Viruses invade the cells; and in the virus diseases that have been suggested to be gene-for-gene diseases the cell nucleus of compatible hosts remains alive.

Jones and Parrott (1965) suggested that the infection of potatoes by the nematode *Heterodera rostochiensis* follows a gene-for-gene pattern. When a larva starts to feed on the cells of the inner cortex and vascular bundle, the cells and their nuclei increase in size. These "giant cells", amalgamated by the loss of some cell walls to form a syncytium, remain alive in susceptible varieties for many weeks (Huijsman et al., 1969).

These few examples show how diverse are the ways in which susceptible host plants respond to infection in a gene-for-gene system. However, despite all this diversity, one feature, on the evidence, remains in common: in all demonstrated gene-for-gene diseases the cell nucleus of susceptible host plants survives infection, sometimes for long, sometimes for short, periods, according to what disease is involved. This simple anatomical relation is the first line of support for the protein-for-protein hypothesis.

The organelles, particularly the mitochondria and chloroplasts, are possibly equally involved in specifying proteins. Stressing the survival of the compatible host cell nucleus is not to be taken as ignoring the cytoplasm. They go together; and when the nucleus survives, so does the cytoplasm. A separate discussion of the cytoplasm is therefore unnecessary.

2.5 The Gene for Susceptibility and Its Different Roles

Biffen (1905), working with stripe rust of wheat caused by *Puccinia striiformis*, showed that susceptibility and resistance to the pathogen segregated in a Mendelian way. For the next 71 years alleles for susceptibilitiy were thought of as bad, and alleles for resistance as good. Vanderplank (1976) saw the matter differently, and argued that the gene for susceptibility was an essential and integral part of the healthy plant, coding a protein necessary for the plant. The role of susceptibility is forced on the gene and its products by the parasite, which turns the protein to its own use. It is basically wrong to think of the gene for susceptibility as a gene for self-inflicted harm to the host.

The gene for susceptibility, for that is what we are forced by circumstances to call it, has at least two roles. It has a secondary role of serving some purpose of the pathogen. This is the role obvious to us that has given the name, gene for susceptibility. It has also a primary role as a normal and essential constituent of the plant, independent of parasitism. This is a role not immediately obvious, and for that reason hitherto overlooked and unnamed.

Consider the wide, often universal, occurrence of genes for susceptibility. *Phytophthora infestans* devastated the potato fields of western Europe in 1845. Those fields were of *Solanum tuberosum* sensu stricto, which had none of the now known *R* genes for resistance the blight. In the 1920s the resistance gene *R*1 from *S. demissum* was used by potato breeders. The progeny of crosses between resistant and susceptible parents segregated roughly 1:1 for resistance: susceptibility. Thus we came to recognize *R*1 and its allele for susceptibility *r*1[1]. All the evidence agrees that *r*1 was universal in *S. tuberosum* until the 1920s when *R*1 was brought

[1] For the genes named in this section, except *sr* 17, resistance is dominant at normal temperatures and susceptibility recessive. The alleles for resistance to potato blight, wheat stem rust, and wheat leaf rust are therefore written *R*, *Sr*, and *Lr*, respectively, and the alleles for susceptibility *r*, *sr*, and *lr*, but this symbolism carries no weight outside resistance/susceptibility phenomena. In their primary function the *r*, *sr*, and *lr* alleles are not necessarily recessive. Nor must they be regarded as poor relations of the *R*, *Sr*, and *Lr* alleles; the special purpose of this sections is to show the opposite, that they are important in their own right.

in. Potatoes were universally susceptible to blight, which they would not have been if *R*1 had been present. We can exclude the possibility that susceptibility in the host resulted from matching virulence in the pathogen, because at that time virulence on *R*1 was so rare that, despite extensive searches by plant pathologists and breeders, it was not discovered until the 1930s.

The universal presence of *r*1 means that it existed for reasons other than to make potatoes susceptible to *P. infestans*. Harmful recessives may occur in a population as a genetic load, but not at a frequency of 100%. The frequency of a harmful allele is maintained by a balance between mutation and selection, from which it follows that an allele universal in a population cannot be just harmful. The allele *r*1 was universal; therefore it was useful.

Wheat and *Puccinia graminis tritici* tell the same story. The great cultivar Marquis dominated the spring wheat areas of Canada and the United States in the early decades of this century. It carried universally and homozygously the alleles for susceptibility *sr*5, *sr*6, *sr*8, *sr*9, *sr*11, *sr*12, *sr*13, *sr*14, *sr*16, *sr*17, and relatively newly recognized alleles such as *sr*24, *sr*25, *sr*26 and *sr*27. All these alleles were unquestionably harmful, in that they allowed *P. graminis tritici* to attack Marquis. Had the corresponding *Sr* alleles been present—even if only one or two of them, such as *Sr*6 and *Sr9d*, had been present—Marquis would have been resistant. The only reasonable conclusion from all this is that the *sr* genes were part of the necessary make-up of Marquis wheat, and had a useful function in Marquis. Why else were they there, at 100% frequency?

What holds for Marquis wheat holds for many other wheat cultivars. The great bulk of the world's wheat has a low ratio of *Sr* to *sr* alleles. At most relevant loci (and there are twenty or more of them) susceptibility is universal in the wheat population, i.e., the *Sr* genes present in any one cultivar number considerably less than twenty.

The situation is perhaps even more marked with the *Lr* and *lr* alleles for resistance or susceptibility to wheat leaf rust caused by *Puccinia recondita*. Although many loci are available, Canadian spring wheat cultivars are mainly monogenic for resistance (Anderson, 1961; Dyck et al., 1966; Samborski and Dyck, 1976) similar patterns are common elsewhere. Alleles for susceptibility are the common alleles, even where *P. recondita* occurs and the threat of disease exists. At most relevant loci, the allele for susceptibility is 100% frequent, despite mutants for resistance being available. All this is incomprehensible except on the concept that the alleles for susceptibility have some role other than to make for susceptibility.

Look at the matter in terms of the process of deletion. The very existence of genes for susceptibility shows that they are needed; otherwise, they would have been deleted if they had ever occurred at all. Deletion is a well-known genetic phenomenon, and higher plants are well equipped to remove unwanted genes. There is among angiosperms at least a 65-fold range in nuclear DNA content in apparently diploid species, i.e., the range is irrespective of polyploidy (Price, 1976). While some of this range is the result of gene duplication (the topic of the next section), much is the result of gene deletion. Within genera a substantial reduction of DNA content may accompany specialization and phylogenic advance (Price, 1976). With efficient mechanisms for deletion, if susceptibility genes were nothing but harmful to the host, why were the loci not deleted? Deletion would have got rid of the *r*, *sr* and *lr* alleles from the host populations of potatoes and wheat and, concomitantly, of the need for *R*, *Sr*, and *Lr* alleles. The loci were not deleted. The *r*, *sr*, and *lr* alleles still exist; evidently they are not just harmful to potatoes and wheat. The exist because potatoes and wheat need them.

There is a twist to the deletion argument that we can put in somewhat anthropomorphized terms. The gene for susceptibility must be essential to the host because the parasite could not risk staking its existence on a host product that the host could jettison by gene deletion, jettisoning the parasite in the process. In intimate parasitism we can take it for granted that the parasite uses at least one host product essential to the host, and the gene specifying that product is necessarily a gene for susceptibility. (The reference to intimate parasitism is in order to exclude, inter alia, scavengers like *Sclerotium rolfsii* which are not gene-for-gene parasites and fall far outside the scope of this chapter.)

A corollary to Vavilov's (1949) rule, apparently missed by Vavilov himself, is that in general the allele for susceptibility is the fitter allele. Vavilov's rule states that resistance to disease is most likely to be found in the gene centers of cultivated plants. The rule has been widely discussed, and the literature has been reviewed by Leppik (1970). The gene center, or center of origin, of a crop plant is the primary region of diversity of cultivated species and varieties. It is the center from which cultivated species and varieties have moved outwards. It is also the center in which disease resistance is most concentrated. The last two sentences, paraphrased, state that as a crop plant moves outwards from its center of origin it takes proportionately more genes for susceptibility with it, leaving proportionately more genes for resistance behind at the center. There is evidently some special virtue to the plant in genes for susceptibility for them to be carried as first choice.

Consider stem and leaf rust of wheat. The origin of the known main gene complexes in cultivated wheats that determine rust resistance can be traced back to the gene center of wild wheats in the Middle East (Leppik, 1970). As wheat spread out, to be cultivated in the Americas, Europe, Australia, Africa, and the Far East, it took with it mainly the *sr* and *lr* alleles, and left behind most of the *Sr* and *Lr* alleles. That is what Vavilov's rule is about. Clearly, the *sr* and *lr* alleles, despite the tag of susceptibility, and despite there being danger from *P. graminis tritici* and *P. recondita* abroad, were useful to the host. This phenomenon of alleles for susceptibility at the periphery and alleles for resistance at the center is no freak of random genetic drift. It is part of a well-established general rule.

In the previous paragraph the argument is not concerned with why resistance abounds in the gene center. The accepted explanation is adequate: an abundance of inoculum provides selection pressure for resistance. The argument is about why, when there was a choice of alleles, it was mainly the *sr*, not the *Sr*, the *lr*, not the *Lr*, alleles that went or were taken out of the gene center. Opportunity for mutation does not enter into the argument. All contemporary wheats, cultivated or wild, inside the gene center or outside, are necessarily of the same age relative to the ancestral wheats, and have had the same time to mutate. Selection pressure does not affect mutation rates; and even if it did, one would still have to face the fact that it was the *sr* and *lr* alleles that were selected, not the *Sr* and *Lr*.

From all directions the evidence converges: what we call genes for susceptibility are in fact genes needed by the host plant for reasons unconnected with the presence or absence of the parasite. They are plant genes, not host plant genes.

In Sections 3.16 and 3.24 the identity of these genes is speculated upon.

2.6 Comments Arising from the Possibility of Tandem Duplication

The short arm of chromosome 10 of maize carries an abundance of genes for resistance. The *Rp*1 locus (or, more correctly, group of loci) has 15 known alleles that determine resistance to rust caused by *Puccinia sorghi*. Near the *Rp*1 locus are two other loci, *Rp*5 and *Rp*6, also determining resistance to *P. sorghi*, and a locus *Rpp*9, determining resistance to *P. polysora*. The evidence has been presented by Saxena and Hooker (1968) and Day (1974).

A similar phenomenon exists with powdery mildew of barley. Most of the resistance genes to *Erysiphe graminis hordei* are on the long arm of chromosome 5 of barley (Wolfe, 1972).

Commenting on the evidence about maize rust, Day (1974) concluded that the region of the short arm of chromosome 10 in maize, carrying the *Rp* and *Rpp* loci has undoubtedly evolved so that it is primarily concerned with the control of rust resistance. This conclusion is based on the old thesis that resistance alleles are good alleles (for the host plant), and susceptibility alleles are bad alleles, a thesis which, for reasons, given in the previous section, we reject entirely in favor of the thesis that what we call genes for susceptibility are necessary constituents of the plant, and have initially nothing to do with parasitism.

What is the alternative? We suggest tandem duplication. That is, we suggest that the necessary genes on chromosome 10 of maize, misnamed genes for susceptibility, were duplicated tandemly, and that some of these duplicates mutated to *Rp* and *Rpp* alleles for resistance. Tandem duplication is a well-recognized phenomenon in eukaryotes; Ohno (1970) has written on a theory that evolution has been a process of gene duplication followed by mutation of redundant duplicates. Our suggestion about tandem duplication is not essential to the protein-for-protein hypothesis, in that the hypothesis does not need evidence that tandem duplication occurs. Nevertheless evidence would be relevant, at least to details, and in relation to the large number of loci involved.

Tandem duplication results in a linear sequence, or several linear sequences, of reiterated genes with noncoding spacers between them. (The word duplication is not restricted to doubling, but refers to any longitudinal repeat of a DNA sequence, irrespective of the extent of the reiteration.)

With genes coding for proteins, repetition, including tandem duplication, may be restricted primarily to those proteins needed in abundance at critical, abnormal times. This is what Price (1976) concluded from a survey of the literature. If Price is correct, this is a valuable pointer to the type of protein involved in protein-for-protein relations. The argument that gene repetition is likely to occur when the gene product is needed in abundance at critical times has also been used by Ohno (1970) for 5S ribosomal RNA; he quotes an estimate that the genome of the frog *Xenopus laevis* contains 20000 duplicated copies of a gene coding for 5S ribosomal RNA.

The evidence, such as it is, for tandem duplication is evidence that many genes are needed. However, the need for many genes is not necessarily reflected in tandem duplication. Diseases vary. In flax the genes involved with rust seem to be concentrated in five groups, called the K, L, M, N and P loci, and of these the N and P loci are linked (Flor, 1959). In wheat the genes involved with stem rust are, however, in 15 groups scattered on at least ten chromosomes, the largest known grouping being on chromosome 2B in a series of six alleles in what is called locus *Sr9* (Daly, 1972, quoting W.Q. Loegering, personal communication).

2.7 The Gene for Avirulence and Its Different Roles

Just as genes for susceptibility have a positive value for the plant, a hidden value not apparent from the name, so too genes for avirulence have a positive value to the parasite, a hidden value not apparent from the name. Much of what was said in Section 2.5 about the usefulness to the plant of what we call genes for susceptibility can be repeated mutatis mutandis about the usefulness to the parasite of what we call genes for avirulence. For avirulence in the parasite the arguments are even more forceful than those for susceptibility in the host. Whereas susceptibility as such is usually recessive, avirulence as such is usually dominant, and can harm a population of the parasite even at low frequencies. Whereas susceptibility is commonly just harmful, the host surviving with varying amounts of injury, avirulence is commonly entirely lethal to the parasite. In relevant incompatible host–parasite combinations, the avirulence gene is commonly a suicide gene.

The adjective, avirulent, must not be confused with nonpathogenic. Avirulence has meaning only with relevance to a particular gene for resistance. Thus a culture of *Phytophthora infestans* avirulent on the resistance gene *R*1 in potatoes can be fully pathogenic on potatoes without the gene *R*1 (or other *R* genes), but dies without reproducing itself (i.e., without forming spores) on potatoes with the gene *R*1. Avirulence on *R*1 is no barrier to infection of potatoes without *R* genes, but is lethal to the parasite on potatoes with *R*1.

Consider from another angle some evidence about *Phytophthora infestans* that has already been presented. In 1845 *P. infestans* struck the potato fields of western Europe. At that time fields of potatoes, being pure *Solanum tuberosum*, had no known *R* genes, and the fungus in Europe did not meet any known *R* genes until the 1920s when the gene *R*1 was experimentally introduced from *S. demissum*. During all the years until the 1920s (and even beyond, because virulence on *R*1 was not detected until 1932) avirulence on *R*1 was universal or nearly universal in the population of *P. infestans*. The evidence for this is that when the gene *R*1 was tested in Europe in the 1920s it gave complete protection against blight. Not a single blight lesion was found in potato lines with the gene *R*1 despite close searching by skeptical plant pathologists and agronomists. This universal occurrence of avirulence must be seen against the background of the rule in population genetics that the frequency of an allele in a population is determined by the balance between mutation and selection. An allele that occurs with 100% frequency cannot therefore be just useless to the population: it must have some positive value. There can be no doubt here about the ability of avirulence to mutate to virulence. Potato breeders who have used resistance from the gene *R*1 know to their cost that, where the climate favors blight, virulence quickly appears.

Also relevant is the famous Fisher (1930) postulate: mutations are recurrent, and all mutations occurring will have done so many times previously in the history of the species. If during the 80 or more years following the introduction of *P. infestans* into Europe, the fungus population carried the avirulence allele with 100% or nearly 100% frequency, it was because this allele has selective value—the fungus needs it—and not because opposite mutations could not arise.

To put the argument in simpler terms, it is unbelievable that an allele would occur with 100% frequency in a population if its only function was at times to be lethal to the phenotype, and doubly improbable that an allele occurring with 100% frequency would be both lethal and dominant, but useless to its possessor.

The high frequency of avirulence is common in many pathogen populations. Indeed, the high frequency of avirulence is a logical deduction from crop breeders' efforts to create resistant cultivars. Every time a wheat breeder purposely introduces a new *Sr* or *Lr* gene, he does so because field experience has shown that the gene introduces resistance against *Puccinia graminis tritici* or *P. recondita;* and resistance in the host is equivalent to avirulence in the pathogen, within the concept of wheat breeding.

The deletion argument is as valid for avirulence in the pathogen as for susceptibility in the host. If avirulence had no role other than at times to harm the phenotype, why was it not deleted during the history of the species? And if it was not removed by deletion, why was it not removed by substitution? The Middle East is the gene center for *Puccinia graminis tritici* (Leppik, 1970). Virulent pathogenic races are abundant there, as are the host plants, *Berberis* spp. and *Mahonia* spp., of the sexual phase. There, if avirulence was nothing but harmful, was *P. graminis tritici's* opportunity to get rid of it, and replace it by virulence.

The other aspect of the gene center argument is as valid for avirulence in the pathogen as for susceptibility in the host. Why if avirulence was useless to the pathogen did avirulence at many loci migrate from the old gene center to newer areas of cultivation? Why was the allele for avirulence the preferred allele to be taken by the pathogen in its migration, the preference being evident from the fact that, until the last 30 or 40 years when plant breeders started using resistance genes, avirulence was far more prevalent that virulence. (Otherwise there would have been no plant breeding programs using *R*, *Sr*, *Lr* and similar resistance genes.)

2.8 Susceptibility and Resistance. Neutral Mutation

Section 2.5 was devoted to establishing the proposition that what are called genes for susceptibility are in fact essential plant genes which operate independently of parasitism. They have a primary role in the plant, and the secondary role which we call susceptibility is something forced on the plant by the parasite. The parasite makes use of a plant gene and turns it into a host plant gene.

What then is the role of a gene for resistance? In the primary role, that is, in the absence of the parasite, it substitutes for the gene for susceptibility. In the absence of *Phytophthora infestans*, what the gene *r*1 does in the potato, that too the gene *R*1 does. In the absence of *Puccinia graminis tritici* what the gene *sr*11 does in wheat that too the gene *Sr*11 does, and so on.

Consider a potato cultivar growing in an environment without *Phytophthora infestans* (because the climate is too dry, or because the plants are fully protected by a fungicide, or because there is no source of inoculum). If an expert on potatoes, who however did not happen to know the particular cultivar, was brought into the field he would not have found it possible to decide whether the cultivar had the gene *r*1 or the gene *R*1. The phenotypes of *r*1 to *R*1 are in the absence of blight undistinguishable. In the absence of blight the mutation is phenotypically neutral.

The whole concept of breeding for resistance to disease assumes that in the absence of the pathogen the change from a gene for susceptibility to a gene for resistance will be phenotypically neutral or nearly neutral. Wheat with, say, *Sr*11 is assumed to be as good agronomically, other things being equal, as a wheat with *sr*11. Breeding for resistance would be a failure if resistance genes made for significant agronomic or horticultural inferiority. In the other direction, breeding for resistance would never have been a major concern of plant pathology if resistance genes had made for significant agronomic or horticultural superiority, because then resistance genes would automatically have been included in cultivars without waiting for the spur of disease.

There is nothing inherently difficult about accepting phenotypical neutrality. If we may briefly anticipate later discussion, we can picture an enzyme to be a protein with (as a realistic illustration) 200 amino acid residues of which ten are in close contact with the substrate for binding and catalysis. Most changes of amino acid would have little or no effect on the catalysis, and in the absence of complicating factors the mutations that produced these changes would be neutral or nearly neutral.

It is in the presence of the pathogen that a gene for susceptibility and a gene for resistance produce such different phenotypes, and live up to their names. Here again there is nothing inherently difficult about accepting this great difference. One might suppose that the enzyme specified by the gene for susceptibility and that specified by the gene for resistance differed by an amino acid not in contact with the substrate for binding or catalysis, but involved in host enzyme–pathogen enzyme copolymerization which will be the central theme of much that follows in this chapter.

2.9 Avirulence and Virulence

Section 2.7 was written to establish the proposition that genes for avirulence have a primary useful role in the pathogen. It is only in the presence of resistance genes in the host that avirulence genes in the pathogen have the secondary role which gives them their name.

What then of a gene for virulence? In the absence of a resistance gene in the host, the virulence gene substitutes for the avirulence gene in its primary role. In the absence of a resistance gene in the host, mutation from avirulence to virulence is neutral or nearly neutral for the phenotype.

Much ink has been spilled on this matter, which will be a major topic of Chapter 6. The issue is this: is there or is there not a stabilizing selection operating against unnecessary virulence, i.e., against virulence when there is no corresponding gene for resistance in the host? Those who dispute that there is stabilizing selection are in effect asserting that mutation from avirulence to virulence is entirely neutral when the relevant resistance gene is absent. Because stabilizing selection, or its absence, is discussed in some detail in Chapter 6, we need only summarize briefly here. Most mutations from avirulence to virulence in the absence of the relevant resistance gene are neutral or very nearly neutral, but a minority, highly relevant to breeding for resistance, are not. This is understandable in terms of enzyme structure. Most amino acid substitutions can be expected to have little or no effect on catalytic activity; but some substitutions, especially those at more critical sites in the molecule, are likely to be significantly detrimental.

2.10 Isozymes

Isozymes are different proteins with distinct genetic origins that catalyze the same reaction. The proteins differ in one or more amino acids, a difference which reflects a difference in the DNA which specifies them.

Some amino acid changes are forbidden, in the sense that they substantially curb or destroy catalytic activity. Amino acid changes that seriously affect the tertiary structure of the proteins, such as those that affect disulfide bridges, are largely forbidden. Other changes are tolerable, with varying tolerability. A change from, say, glycine to alanine would scarcely be noticed catalytically, if it was sited away from that part of the protein bound to the substrate. The mutation would be neutral.

By definition, isozymes catalyze the same reaction. Their effect may vary *quantitatively*, that is, one isozyme may be more efficient catalytically than another. The product, however, stays the same, that is, there is no *qualitative* difference in the product.

Isozymes are the last great reservoir of genetically controlled qualitative differences; this will be remembered from Section 2.2. Of genetic variation involving base substitution some stops at the RNA. This is samesense mutation, which changes the DNA and RNA but not the specified amino acid. Of the variation that reaches enzymes, most occurs as isozymes. Isozyme variation, in terms of catalysis but not polymerization, stops qualitatively with the enzyme, and does not reach the product of the catalysis.

Gene-for-gene relations are concerned with differential interactions, that is, with associated variation of host and pathogen. Variation in the host, in the form of a qualitative change from susceptibility to resistance, can be countered by a variation in the pathogen, in the form of a qualitative change from avirulence to virulence. There must be variation, or potential variation, in *both* host and pathogen. Because qualitative variation in isozymes is not passed on catalytically, it is reasonable to suggest that it is not through catalysis but through polymerization that associated variation of host and parasite (i.e., in *both* partners) is conserved (see Chap. 9 for more detail).

The activity of proteins as catalysts and as polymerizers has different origins. The amino acid residues that govern catalysis are to be found at or near the site of binding with the substrate, and many have hydrophilic side chains. The amino acid residues that govern polymerization owe their activity to hydrophobicity, and many have hydrophobic side chains. Because of these differences, it is possible for isozymes to be neutral in catalysis bit distinguishable in copolymerization. This dualism explains the dualism in the previous few sections.

2.11 The Hydrophobic Effect. Amino Acids with Hydrophobic Side Chains. Their Interrelated Genetic Codes

Consider a protein in solution in water. If the molecular units stay separate in the water as monomers, each monomer being surrounded by water, it is because they have more affinity for water than for one another. The surface of the monomers is hydrophilic. But if two or more units came together, burying parts of their surface in one another, they are rejecting water in favor of protein. The buried surfaces are hydrophobic.

Hydrophobic binding is the basis of protein polymerization in water and of the protein-for-protein hypothesis of this chapter. This chapter is concerned with evidence about a group of diseases, which for convenience we call gene-for-gene diseases, in which, we hypothesize, pathogen and host recognize each other because of the hydrophobic binding involved in protein copolymerization.

For our purpose we shall define a monomer as the largest polypeptide unit held together by covalent bonds. Polymerization, then, involves only noncovalent bonding[2]; the hydrophobic binding is noncovalent. It is noncovalent binding that gives protein polymers the great thermodynamic flexibility absent from covalently bound polymers. Indeed, most of the next chapter is concerned with mat-

[2] There are also covalently bound protein polymers. Enzymes that catalyze sequential or related reactions are often found as multienzyme complexes bound covalently in a single polypeptide chain. This has been reviewed by Kirschner and Bisswanger (1976). The original rule, one gene–one enzyme, is more correctly stated as one gene–one polypeptide which may be multienzymatic. Covalent binding of this sort is possible only when the protein is specified by a single genome, and is excluded from copolymers in which some subunits come from the host and some from the pathogen. It is therefore excluded from our discussions.

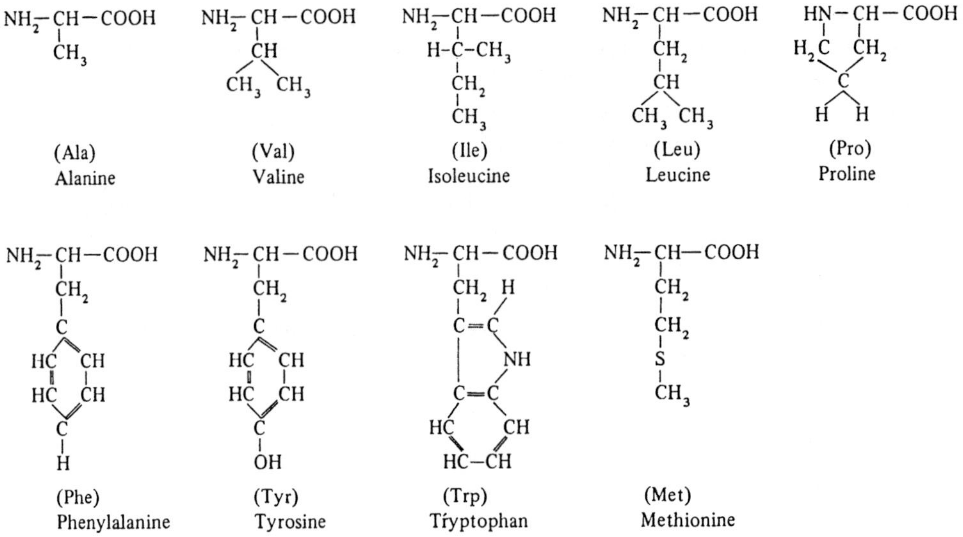

(Ala) Alanine | (Val) Valine | (Ile) Isoleucine | (Leu) Leucine | (Pro) Proline

(Phe) Phenylalanine | (Tyr) Tyrosine | (Trp) Tryptophan | (Met) Methionine

Fig. 2.1. Amino acids with hydrophobic side chains

ters that derive directly from noncovalency of protein polymers. Also, noncovalent binding allows a protein copolymer to be specified by two separate genomes, of host and parasite; and this is the foundation on which the whole of our hypothesis is built.

The disrupting and distorting of the relatively strong bonds between water molecules are as much a part of the hydrophobic effect as the weak bonds between hydrocarbon side chains themselves. That is, the hydrocarbon side chains come together as a sort of "oil droplet", not only because of internal hydrocarbon affinity but also because of the internal affinity of water molecules for one another. The relative weakness of bonds between hydrocarbon molecules can be seen by comparing the simplest hydrocarbon, methane CH_4, with water. The molecular masses of methane and water, 16 and 18, are roughly the same. Yet methane boils at —160° C and water at 100° C. This difference of 260° C reflects the relatively strong bonds that hold H_2O molecules together in liquid water and must be broken when water vaporizes. An H_2O molecule in water as liquid can make four hydrogen bonds with neighboring H_2O molecules. These are the bonds that must be broken or bent when protein is dissolved in water, i.e., when protein molecules make "holes" in water. All this does not affect the definition of protein polymerization as a hydrophobic effect, nor does it lessen the importance of side chains being hydrophobic. Hydrophilic side chains can themselves form bonds with water, thereby compensating for the disruption and distortion of the bonds that exist in pure water. This, for all practical purposes, hydrophobic side chains are unable to do; and from this inability stems much of the hydrophobic effect.

An account of the thermodynamics of protein polymerization is given by Oosawa and Asakura (1975).

(Ser) Serine	(Thr) Threonine	(Cys) Cysteine	(Asp) Aspartic Acid	(Glu) Glutamic Acid
H_2–CH–COOH | CH_2 | OH	NH_2–CH–COOH | H-C–OH | CH_3	NH_2–CH–COOH | CH_2 | SH	NH_2–CH–COOH | CH_2 | COOH	NH_2–CH–COOH | CH_2 | CH_2 | COOH

(Lys) Lysine	(Arg) Arginine	(Asn) Asparagine	(Gln) Glutamine	(His) Histidine
H_2–CH–COOH | CH_2 | CH_2 | CH_2 | CH_2 | NH_2	NH_2–CH–COOH | CH_2 | CH_2 | CH_2 | NH | C=NH | NH_2	NH_2–CH–COOH | CH_2 | C=O | NH_2	NH_2–CH–COOH | CH_2 | CH_2 | C=O | NH_2	NH_2–CH–COOH | CH_2 | C=CH | | N NH \\ / C | H

Fig. 2.2. Amino acids with hydrophilic side chains

Of the 20 essential amino acids nine have hydrophobic side chains. These are shown with their conventional abbreviations in Figure 2.1. In all except three, the side chain is pure hydrocarbon. The three are tyrosine, with a phenolic OH group that can ionize to some extent but not sufficiently to cancel the hydrophobicity of the rest of the side chain, tryptophan (the most hydrophobic of all the essential amino acids) with a not very active NH group, and methionine, with a locked-up S atom not nearly as active as the S atom in cysteine. In addition to these nine amino acids there are others with hydrophobic moieties in what are on balance strongly hydrophilic side chains, such as the $(CH_2)4$ part of the lysine side chain (Fig. 2.2).

Table 2.3 shows the genetic code of the nine amino acids with hydrophobic side chains. It is constructed on conventional lines to be found in reference works on molecular biology, and these should be consulted for details. What the table brings out is the close interrelation of the codes for many of the amino acids with hydrophobic side chains. Thus every amino acid with U as the middle letter of its coding triplet has a hydrophobic side chain, so that mutations which change the first or third letter leave the side chain hydrophobic. Even allowing for the fact that the hydrophobic side chains differ greatly in hydrophobicity (with the CH_3 of alanine by far the weakest), this interrelation makes for considerable stability in the face of mutation. Even more stabilizing is the high proportion of samesense mutations, i.e., mutations that affect the DNA and RNA but not the coded amino acid. There are six different triplets all specifying leucine. The exceptions to the interrelating codes are those for tyrosine and tryptophan, both with strongly

Table 2.3. The genetic code for amino acids with hydrophobic side chains, each amino acid being specified by a coding triplet in messenger RNA

First letter	Middle letter				Last letter
	U	C	A	G	
U	Phe		Tyr		U
	Phe		Tyr		C
	Leu				A
	Leu			Trp	G
C	Leu	Pro			U
	Leu	Pro			C
	Leu	Pro			A
	Leu	Pro			G
A	Ile				U
	Ile				C
	Ile				A
	Met				G
G	Val	Ala			U
	Val	Ala			C
	Val	Ala			A
	Val	Ala			G

hydrophobic side chains. Most mutations affecting any of the three letters of the triplets specifying these two amino acids are likely to affect the protein molecule strongly.

2.12 Amino Acids with Hydrophilic Side Chains. Their Part in Catalysis

The amino acids with hydrophilic side chains help to stabilize protein structure, the topic of the next section, and determine protein specificity, the topic of Section 2.14. They cannot be left out of a balanced view of our problem.

The ten essential amino acids with hydrophilic side chains are listed in Figure 2.2. Considering side chains only, two of these amino acids are alcohols with hydroxy OH, two carboxylic acids, two amines, two amides, one has a thiol — SH group, and histidine has an imidazole ring. These substitutions in an otherwise hydrocarbon side chain increase solubility in water. This can be illustrated by comparing the solubility in water of the hydrocarbon methane CH_4 with that of its relevant substitution products. Methane is only slightly soluble; methyl alcohol $CH_3{\cdot}OH$, formic acid HCOOH, and formamide $HCO{\cdot}NH_2$ are liquids either miscible with water or very soluble in it; methyl amine $CH_3{\cdot}NH_2$ is a gas about 20000 times as soluble as methane itself; and the thiol derivative $CH_3{\cdot}SH$ (mercaptan) reacts with water and decomposes.

Considered broadly, the same reactivity that makes a side chain more soluble in water makes it more reactive in catalysis. Hydrophobicity and ability to catalyze are, however, not entirely opposites. Tyrosine for example has a hydrophobic side chain that nevertheless takes part in catalysis by virtue of being a phenol. All the same, it is an approximate generalization to say that enzyme polymers need at least two sorts of surface areas, with hydrophobic side chains to promote polymerization and hydrophilic side chains to promote catalysis.

2.13 Protein Structures

The essential amino acids, with the exception in detail of proline, have the structure

$$NH_2—CH{\cdot}R—COOH$$

where R is the side chain. They combine by forming a peptide linkage

$$—CO—NH—$$

in a polypeptide chain

$$—NH—CH{\cdot}R_1—CO—NH—CH{\cdot}R_2—CO—$$

The term, amino acid residue, is used for

$$—NH—CH \cdot R—CO—$$

and a polypeptide chain typically contains 100 to 200 or more amino acid residues.

The primary structure of a protein is the linear sequence of amino acid residues in the polypeptide chain. The sequence is fixed for each protein. In turn, the sequence of amino acid residues in the polypeptide chain is determined by the sequence of coding triplets (and hence of the nucleotides themselves) in the DNA of the gene concerned. The sequence of amino acids is at protein level what the sequence of coding triplets is at genetic level. Further, the sequence of amino acid residues in the polypeptide chain determines all the higher protein structures. A given primary structure determines the secondary, tertiary, and quaternary structure.

The secondary protein structure, important in fibers like silk and wool, does not fall into our discussion.

In the tertiary structure the polypeptide chain folds into the globular form which enzymes have. The whole structure is held together by noncovalent bonds, particularly hydrogen bonds. The folding brings together on the surface amino acid residues which may be far apart on the primary polypeptide chain. The hydrophilic side chains are probably all orientated so as to remain in contact with the surrounding water. However, the converse is not true; all hydrophobic side chains are not buried inside the protein molecule. A high proportion of hydrophobic side chains lie at the surface in contact with water. This is true even for the purely hydrocarbon side chains such as those of alanine or valine, and especially true of the aromatics, phenylalanine and tryptophan (Klotz, 1970). This probably follows from the sequence of amino acid residues in the primary polypeptide chain. The hydrophobic residues are not grouped together in sequence, nor are the hydrophilic residues. They are interspersed in such a way that internal structure and energy relations leave hydrophobic residues at the surface.

There is another way of looking at why hydrophobic side chains appear at the surface of the molecule in contact with water, despite their hydrophobicity. As a rule, proteins that contain less than 30% of amino acid residues with hydrophobic side chains exist as monomers in water. When the proportion exceeds 30%, there is a tendency for the proteins to polymerize in water (Van Holde, 1966). When they make up less than 30% there is space for the hydrophobic side chains to be orientated inwards, away from contact with water, and for the hydrophilic side chains to be orientated outwards, in contact with water. The hydrophobicity at the surface is then too low for polymerization; but with high contents of hydrophobic amino acid residues, many hydrophobic side chains are for internal space reasons necessarily outwardly orientated and available for hydrophobic interactions.

Every protein has a unique surface pattern that traces back to the sequence and interspersion of residues on the primary chain; and a relevant feature is that the areas which associate when monomers come together as polymers are somewhat more hydrophobic than the surface as a whole (Chothia and Janin, 1975).

A quarternary structure results from polymerization. Structural analyses have shown that protein polymers in general have regular structures; and Oosawa and Asakura (1975) have discussed both the polymerization process leading to a particular structure and the polymorphic transition from one structure to another. Quaternary structure varies greatly with the protein, and is characteristic of the protein. Bacterial flagella are head-to-tail polymers of large numbers of molecules, especially of the protein flagellin. Normal hemoglobin is a tetramer which is a tetrahedron. The quaternary form is sensitive to mutation and change. The difference of a single amino acid residue in flagellin molecules results in two different shapes of flagella, normal and curly. The change of a single amino acid residue, from that of glutamic acid to that of valine, in a half molecule of 287 amino acid residues, changes normal tetrahedral hemoglobin to sickle cell hemoglobin which is a tubular polymer.

2.14 Protein–Protein Recognition

If protein monomers had large patches on the surface consisting of nothing but amino acid residues with hydrocarbon side chains, the monomers would come together as polymers, with patch associated with patch; but such polymerization would be unspecific. Any monomer would associate with any other monomer, randomly. Such polymerization would contribute nothing towards knowing how a parasite recognizes a host. For parasite–host recognition at the chemical level there must be a recognition which in terms of our protein-for-protein hypothesis is a parasite protein–host protein recognition. There must be a contact between protein surfaces in such a way that one surface recognizes another as the one and only possible partner with which it will associate.

Chothia and Janin (1975) investigated protein–protein recognition. Hydrophobicity, they found, is the major factor stabilizing protein–protein associations; but the contribution of hydrophobicity is entirely unspecific. Specificity requires that the surfaces involved must complement each other spatially, and the polar atoms must be properly positioned to make hydrogen bonds between the surfaces.

Consider this in more detail. Specificity involves both an attraction and a spatial arrangement. All atoms or molecules attract one another weakly. This attraction is the van der Waal's attraction. (It is the electrostatic attraction of the nuclei of one molecule for the electrons of another, less the repulsion of electrons by electrons and nuclei by nuclei.) The van der Waal's attractions are numerous and spread over the whole interface, but they are unspecific. Hydrophobic bonding is also unspecific. The attraction which imparts specificity directly is by hydrogen bonds. Hydrogen bonds occur between weakly acidic hydrogen atoms and (in terms of our topic) N, O, and S atoms. These bonds, unlike the van der Waal's bonds, are spatially restricted. They require that the N, O, or S atoms can share a hydrogen nucleus. To act effectively across the interface the N, O, and S atoms must be positioned so that the distance between them is less than 3.7Å; the less the distance, the stronger the action. The N, O, and S atoms are (with the exceptions discussed earlier) in the hydrophilic, not the hydrophobic, side chains, and, in so far as hydrogen bonding determines protein–protein recognition, it is the amino acid residues with hydrophilic side chains that are the more important. The spatial arrangement affecting recognition is the closeness with which the two associating surfaces can approach each other. For recognition, the surfaces must pack closely; they must be so contoured as to complement each other and fit together. Close fitting increases the hydrophobic effect and allows the van der Waal's and hydrogen bonds to act more strongly. It has a specific recognition effect, imparted by the complementarity of contours; and this effect occurs even when the bonds drawing the surfaces together are entirely unspecific.

Chothia and Janin studied the insulin dimer (among other polymers). There are 16 hydrogen bonds at the interface of this dimer, with the N, O, and S atoms so positioned that the distances of bonding are less than 3.7Å. The interface area is approximately 600$Å^2$. These figures show how greatly the positioning of the bonds can contribute to surface–surface recognition.

We can think of protein–protein recognition in terms of fingerprints. Each of us has a unique fingerprint. So too a protein surface, because of its contours and positioning of N, O, and S atoms, is unique. It can recognize another equally unique surface, with complementary contours and positioning of N, O, and S atoms; and, if it is hydrophobic enough, associate with it.

In complementary protein fingerprints we see, at molecular level, the recognition of host by parasite according to Flor's gene-for-gene hypothesis. For every gene conditioning resistance in the host plant there is a specific gene conditioning virulence in the parasite. The protein specified by the gene for resistance in the host plant has a fingerprint that is recognized by the complementary fingerprint on the protein specified by the corresponding gene for virulence in the pathogen. The protein-for-protein hypothesis has a wider compass than the gene-for-gene hypothesis. In the protein-for-protein hypothesis, recognition extends also to the proteins involved in the combination of a gene for susceptibility in the host and a gene for avirulence in the parasite, and in the combination of a gene for susceptibility in the host and a gene for virulence in the parasite.

The general problem in host–parasite specificity at molecular level is to determine which parasite molecules and which host molecules can recognize one another. Recognition implies molecular contact, at least over a distance within the range of chemical bonding. This recognition must be almost limitless in its variety; it must be able to cope with potentially millions of races of cereal rusts, thousands of races of potato blight, and so on. Protein fingerprinting can do this, if one may judge by protein recognition in antibody–antigen systems in animals.

Antibodies are proteins. Antigens are commonly (but not always) proteins. Antibodies and antigens recognize each other, specifically and in endless variety, through protein–protein recognition, i.e., through complementary fingerprinting. The same basic process of protein–protein recognition is used in both the animal and the plant world, with this difference. Animal hosts use protein–protein recognition to inactivate the pathogen, i.e., as a means of defense. Plant pathogens, of the gene-for-gene sort, use protein–protein recognition to draw food from the host, i.e., as a means of attack. The common feature is the almost endless variety of recognition permitted by surface complementarity. The marvel of specific recognition that is commonplace in serology is no more of a marvel than the specific recognition of parasite and host in a gene-for-gene system because, according to our hypothesis, it is the same marvel.

Chapter 3 The Protein-for-Protein Hypothesis: Temperature Effects and Other Matters

3.1 Introduction

The previous chapter sketched some necessary background, but it was noncommittal on an essential question: does protein copolymerization go with the host–pathogen compatibility, which means host susceptibility, or does it go with host–pathogen incompatibility, which means host resistance? This chapter answers the question unequivocally: copolymerization and susceptibility go together.

The evidence comes from various sources. Mainly it comes from a very simple source, the accumulated evidence over the years about the effect of temperature on resistance. Most of this evidence was published in relation to studies of the effect of temperature on reaction types, and hence on the identification of pathogenic races. Fortunately for our purpose, these studies have been popular for the past half century, and have left a mass of data ready to hand for thermodynamic analysis.

3.2 The Thermodynamic Problem

Protein polymerization at relevant temperatures is a reversible endothermic process and, by Le Chatelier's principle, proceeds further at higher temperatures. Two illustrations will suffice.

Lauffer et al. (1958) showed that when tobacco mosaic virus protein in solution with a 0.1 ionic strength phosphate buffer at pH 6.5 was held at laboratory temperature it polymerized into rods, but when cooled to near 0° C it disaggregated into monomers. Also, electron micrographs invariably showed no rods when the protein was sprayed on the grids at 4° C, but always showed rods of the same diameter as tobacco mosaic virus when the grids were sprayed and dried at 25° C.

The second example concerns the mechanism of mitosis. The mitotic spindle to which chromosomes are attached is a head-to-tail polymer, mainly of the protein tubulin. The protein is in the form of fibers when it is polymerized; the fiber structure depends on polymerization, and breaks down when the polymer disintegrates. The parallel arrangement of the polymer fibers makes the spindle birefringent; and the amount of spindle in the cell, and hence the amount of polymerization, can be studied by a polarizing microscope. Inoué (1959), working with the pollen mother cell of *Lilium longiflorum* and with other organisms, found that the spindle is in a temperature sensitive equilibrium. At a temperature of 4–

6° C the spindle birefringence is abolished completely. When the temperature is raised, birefringence returns. The process is reversible, and nuclear division proceeds only at temperatures showing birefringence.

The temperature effect has been widely studied in various contexts (Oosawa and Asakura, 1975), and is consistent. Higher temperatures enhance polymerization; lower temperatures decrease it.

Protein polymerization is a reversible process, which, in terms of a dimer, can be written

monomer A + monomer B ⇄ dimer

At any temperature an equilibrium is reached between association into polymers and dissociation into monomers. The higher the temperature within normal limits, the greater is the tendency to associate. The behavior of any particular protein depends on its tendency to associate. If it associates strongly, i.e., if the hydrophobic effect is strong, polymers will be stable even at low temperature. If the hydrophobic effect is weak, there may be little or no polymerization even at high temperatures. Tobacco mosaic virus protein and mitotic spindle protein are intermediate, and conveniently illustrate a sharp change over biologically relevant temperatures.

This reference to biologically relevant temperatures must be remembered always. Thermodynamics deals with temperatures anywhere on the Kelvin scale, but in the biology of plant disease it can be applied only at temperatures relevant to disease, i.e., at temperatures ranging from the minimum cardinal temperature for disease at the lower limit to the maximum cardinal temperature at the upper limit. We are perforce restricted to considering changes in resistance or susceptibility within these limits. Fortunately for our study, changes within these limits are not rare, and supply ample material for analysis.

The change, if there is a change, is almost invariably in the direction of a loss of resistance as temperature increases, and this connects susceptibility with polymerization. The very few known exceptions can probably be ascribed to monomer inactivation, the topic of the next section.

3.3 Inactivation of Protein Monomers

At high temperatures the tertiary structure of proteins becomes unstable, with a breaking of the internal noncovalent bonds (hydrogen bonds, disulfide bridges, etc.). The conformational change with temperature can be followed by dilatometry and optical measurements (Oosawa and Asakura, 1975). This change inactivates the monomers and prevents them from polymerizing. With the protein flagellin (one of the proteins in the flagella of bacteria) inactivation by heat begins at 28° C and is complete at 50° C (Oosawa and Asakura, 1975).

At room or field temperatures proteins (with the necessary hydrophobicity) polymerize, and polymerization is endothermic. At higher temperatures, when monomer inactivation begins, the polymers begin to break down, and polymerization decreases with increasing temperature.

Except briefly, in Section 3.8, we ignore the inactivation of proteins at high temperatures. As a matter of experimental observation, susceptibility is endothermic; this is what most of this chapter is about. Reversals are rare, and found only at high temperatures well above the cardinal optimum temperature for disease. Also, apart from the experimental evidence, we can probably accept as a matter of principle that substantial inactivation of enzymes will not occur at the normal temperature ranges at which organisms exist. Both in fact and in theory we associate monomer inactivation mainly with artifacts, as when infected plants are held in cabinets and greenhouses at temperatures well above the averages at which natural epidemics occur.

The temperature, 28° C, at which the inactivation of flagellin monomers begins, does not necessarily apply to all proteins—there is practically no evidence about the variability of proteins here—but it is worth remembering that there is a ceiling, whatever it may be, to endothermic polymerization. At present it is only with *Xanthomonas malvacearum* in cotton and *Puccinia recondita* in wheat that there is evidence for a reversal, and then only at temperatures much higher than the optimum for disease. This matter is taken up again in Section 3.7.

3.4 Effect of Temperature on Resistance to Wheat and Oat Stem Rust and Oat Crown Rust

We return here to temperatures likely to prevail in fields in which disease can occur in a susceptible variety, given the necessary inoculum and environmental conditions such as atmospheric humidity.

At 25° C wheat plants homozygous for the gene *Sr6* are susceptible to cultures of *Puccinia graminis tritici* to which they are resistant at 20° C. That is, as a resistance gene, gene *Sr6* loses its effectiveness between 20 and 25° C; between 20 and 25° C resistance changes to susceptibility. To relate this to our hypothesis, i.e., to relate susceptibility with an endothermic process, it will be shown that *P. graminis tritici*, at relevant temperatures, is not thermophilic, and that the change to susceptibility cannot be ascribed to a change of optimal temperatures. In susceptible varieties of wheat *P. graminis tritici*, it will be shown, is at the peak of its aggressiveness at temperatures at which the gene *Sr6* provides resistance, and its aggressiveness is already on the wane at temperatures at which resistance fails. In terms of temperature, the loss of effectiveness of the gene *Sr6* cannot be ascribed to special aggressiveness of the pathogen.

Consider first the effect of temperature on disease in wheat varieties that are susceptible to stem rust, i.e., that lack a relevant *Sr* gene. Stakman and Levine (1919) and Johnson and Newton (1937) concluded that the optimal temperature for wheat stem rust infection lay between 19 and 21° C. Katsuya and Green (1967) inoculated three susceptible wheat varieties (Red Bobs, Little Club, and Marquis) with a suspension of uredospores (40 mg/100 ml) of *P. graminis tritici*. With race 15B-1 of this pathogen the resulting number of pustules per leaf, averaged for the three wheat varieties, was 10.6 at 15° C, 13.9 at 20° C, and 8.4 at 25° C. With race

56 the corresponding averages were 17.4, 20.4, and 12.3, respectively. Intrapolation puts the optimum temperature at a little below 20° C and the most relevant point is that the aggressiveness of the pathogen was clearly on the wane at 25° C the temperature at which the gene *Sr*6 becomes ineffective. Field studies are equally apt. Levine (1928) analyzed data on three hard red spring wheat varieties from 1919 through 1923. The varieties were in uniform rust nurseries in Texas, Kansas, Nebraska, Iowa, Colorado, Wisconsin, and Minnesota. The correlation was based on 264 varietal units in 89 nursery years. The temperature he used in his analyses was the mean temperature during the last two months of wheat growth (which is roughly the temperature a month before ripening, and therefore highly relevant to stem rust studies). The correlation showed that 20° C was about optimal for rust to occur, and there was a rapid decrease in the incidence of rust below 17 or above 22° C. Another analysis was made by Stakman and Lambert (1928). They used temperature data from 1904 through 1925, for Minnesota, North Dakota, and South Dakota. Their criterion was the average temperature for May, June, and July. The inclusion of May might surprise modern workers, with a different background. The years analyzed were before barberry bushes had been eradicated. Stakman and Lambert explain that during May, June, and July *Puccinia graminis* spreads from barberry bushes and multiplies in the uredinal stage on wheat. After the first week in August, wheat, they explain, is usually mature or nearly so, and rust does not develop further. The average temperature of the three months over the 22-year period was 17° C. 1916 was the year of greatest losses from stem rust; in that year it was estimated that stem rust caused a loss of 185 million bushels of wheat (Stakman and Fletcher, 1930) in the barberry eradication States. The average temperature of May, June, and July was about average, i.e., 17° C. The year of the second greatest loss was 1904, when the temperature was below average. The purpose of referring to these years is not to dispute Stakman and Lambert's conclusion that in general the warmer years were the years of more stem rust. The purpose is solely to demonstrate that *Puccinia graminis* can be devastatingly destructive and therefore, by inference, fully aggressive at temperatures well within the range at which the gene *Sr*6 provides resistance.

The effect of temperature on the effectiveness of *Sr*6 as a resistance gene has nothing to do with the effect of temperature on the aggressiveness of the pathogen. This gene is most effective against the pathogen when the pathogen is at the peak of its aggressiveness, as judged by its behavior on susceptible varieties without *Sr*6, and ineffective at the higher temperatures that weaken the pathogen.

Can we then (to reverse our thoughts) assume that a peak of aggressiveness in the pathogen is needed to stimulate the resistance mechanism? The answer is, no. Resistance persists even when the aggressiveness is weakened by suboptimal temperatures; it persists (so far as is known) even when the temperature is reduced to the cardinal minimum, and aggressiveness vanishes. Can the failure of *Sr*6 to condition resistance above 25° C be ascribed to a weakening of host plant? Here again the answer is, no. Wheat plants suffer no apparent injury when held at 36° C, the highest temperature tested by Johnson and Newton (1937), provided that the soil moisture is adequate.

Let us leave the gene *Sr* 6 for the moment, and widen the scope of inquiry. Waterhouse (1929) observed that the reaction on several durum wheat lines to *P. graminis tritici* in the greenhouse varied from resistant in the winter when temperatures were low and light conditions poor to susceptible in the summer when temperatures were higher and light brighter. The same change with temperature and light occurred with race 1 of *P. graminis avenae* on Joanette oats. Johnson (1931) found several durum wheat varieties to be resistant to various races of *P. graminis tritici* at 16° C, but susceptible to the same races at 24° C. Mahomet (1954) found that race 139 of *P. graminis tritici* gave a type "2" reaction on Marquis and Kota wheats at 16–22° C, but a type "3" or "4" reaction at 29–32° C. Daly (1949) found that Thatcher wheat grown with ammonium sulphate as a source of nitrogen was resistant to race 56 of *P. graminis tritici* at 18–24° C, but susceptible above 27° C. When nitrogen was supplied as nitrate, the change was from mesothetic to susceptible over the same range of temperature. Bromfield (1961) found that of the varieties he tested 20 inoculated with race 17, 12 with race 38, and 15 with race 56 were resistant at 21° C, but susceptible at 25–29° C. Watson and Luig (1966) observed that the gene *Sr*15 behaved somewhat like *Sr*6; it is effective at temperatures up to 18–21° C, but not above. They also observed that the variety Celebration, with resistance from Marquillo, lost its resistance to race 21 at 24 to 27° C. Gough and Merkle (1971) found that two genes from the variety Agent for resistance to stem rust race 111-SS2 and two from Agrus, including the gene from *Agropyron elongatum*, were effective at 20° C, moderately effective at 25° C, but ineffective at 30° C. Luig and Rajaram (1972) found that the gene *Sr*5 in a Marquis background was effective against stem rust race 21-4,5 at temperatures up to 24° C but not at 30° C. So too the genes *Sr*8 and *Sr9b*, when heterozygous, became ineffective against this race at high temperatures. Sanghi and Luig (1974) found the gene *SrFr*2 to be temperature-sensitive; it becomes ineffective at 27° C.

To summarize, the literature about wheat stem rust is consistent: If temperature affects resistance, it affects it in the direction of reduced resistance at higher temperatures.

Gordon (1933) confirmed and extended the finding of Waterhouse with Joanette oats. At 12° C the variety was resistant to races 1, 3, 4, and 5 of *P. graminis avenae*. At 24° C it was fully susceptible to races 3 and 4, and moderately susceptible to races 1 and 5. At 28° C it was fully susceptible to all four races. Newton and Johnson (1944) confirmed these results with stem rust on Joanette oats, and added another example of the breakdown of resistance at high temperatures: Some Hagira crosses, resistant at low temperatures, became susceptible at 27° C. Also, the variety Sevenothree, which is normally resistant, often gave a mesothetic reaction in the heat of summer. Hingorani (1947), quoted by Hart (1949), found that Jostrain oats were intermediate in resistance (infection type "X"=) to race 2 of *P. graminis avenae* at 16° C but fully susceptible at 21–24° C. This same oat variety was moderately resistant (infection type "X"−) to races 5 and 10 at 18° C, more susceptible (infection types "X" or "X"+) at 21–24° C, and susceptible at 29° C. Ibrahim (1949), quoted by Hart (1949), found that the oat variety Garry, and certain selections having the Hagira type of stem rust resistance were resistant to race 6 of *P. graminis avenae* under ordinary conditions of growth, but susceptible at 29° C. Roberts and Moore (1956) found four varieties of oats to be resistant to stem rust at 24° C, but susceptible at 29° C. Martens et al. (1967) studied the stem rust resistance genes in oats individually. Of the five resistance genes available, only genes A and D conferred resistance that was stable up to the highest temperature tested, 30° C. Gene H conferred resistance effective up to 25° C, but ineffective at 30° C. Gene F conferred resistance effective at 20° C, variably effective at 25° C, but ineffective at 30° C. Genes B and E conferred resistance effective at 20° C, but not at 25 or 30° C.

To turn to crown rust of oats, Peturson (1930) found that the variety Red Rustproof was moderately resistant to race 4 of *P. coronata avenae* at 14° C, but highly susceptible at 21–25° C. The varieties Green Mountain, White Tartar, and Green Russian were resistant to race 7 at 21° C, but susceptible at 25° C. Rosen (1955) found that selections from a mutant of Tennessee 1922 x Bond-Ingold cross were resistant to races 45, 57, and 101 of crown rust at 16° C, but susceptible at 21–24° C. For some data of Simons (1954) see Section 3.10, Table 3.6. Zimmer and Schafer (1961) found the oat variety Glabrota to be resistant to race 263 of crown rust at 16° C, intermediate at 21° C, and susceptible at 27° C.

Mesothetic or intermediate reactions keep recurring in the literature of temperature effects, a point made by Johnson (1931). He found that any isolate of *P. graminis tritici* that produced an intermediate ("X") type of infection on durum wheats at ordinary greenhouse temperatures was likely to produce a susceptible ("4") type at higher and a resistant ("1"–"0") type at lower temperatures. Gordon (1930) demonstrated a similar behavior for *P. graminis avenae* and Peturson (1930) for *P. coronata avenae* when forms that produce an "X" type of infection were used. Other examples have been given by Johnson and Schafer (1965) for wheat leaf rust, and Luig and Watson (1965) and Luig and Rajaram (1972) for wheat stem rust. A mesothetic "X" infection type can be given by the gene *Sr* 5 at 27–30° C, by *Sr* 6 at 18° C, and by *Sr* 15 at 15° C. These are roughly the temperatures transitional from effectiveness to ineffectiveness, i.e., from resistance to susceptibility, for the genes concerned.

This sensitivity of intermediate types of reaction is predictable. If full susceptibility means protein copolymerization, with the pathogen's contribution mostly locked up in a copolymer, and if full resistance means lack of copolymerization, or weak copolymerization, with the pathogen's contribution mostly free to act as elicitor of incompatibility, one can expect an intermediate stage of balance between the extremes. This balance will be more easily tipped one way or the other with intermediates than with either strongly associated or strongly dissociated proteins. For the same reason, with an intermediate stage of balance, one would expect greater variability of reaction from one group of cells to another, depending on variability of sugar content or other factors affecting polymerization; and a variable reaction of the cells is what a mesothetic response signifies.

3.5 Temperature and Infection of Tomato Plants by Tobacco Mosaic Virus

Cirulli and Ciccarese (1975) studied the effect of temperature on infection of tomato plants by tobacco mosaic virus, and their results accord in every detail with our hypothesis.

The combination, tomato-tobacco mosaic virus, is thought to be on a gene-for-gene basis (Pelham, 1966). Cirulli and Ciccarese used tomato plants carrying the resistance genes *Tm* 1, *Tm* 2, and *Tm* 2 *a*, singly and in combination. They tested them for resistance against isolates of tobacco mosaic virus collected from field-grown plants of tomato, tobacco, pepper (*Capsicum annuum*) and *Solanum nigrum*.

Higher temperatures reduced resistance. Thus, plants heterozygous for the gene *Tm* 1 were resistant to 42% of the virus isolates at 17° C, to 26% at 22° C, to 5% at 26° C, and to none at 30° C. A broader view of the temperature effect is given in Table 3.1 which gives data averaged for the three *Tm* genes. The effect of temperature is consistent, increasing temperature regularly reducing resistance as expressed in terms of the percentage of avirulent isolates of tobacco mosaic virus.

The loss of resistance resulting from higher temperatures is spread over the whole range of temperature from 17–30° C (see Table 3.2; for example, plants heterozygous for *Tm* 1 were resistant to 42% of the isolates at 17° C and 26% at

Table 3.1. Effect of temperature on the percentage of isolates of tobacco mosaic virus against which the tomato resistance genes *Tm*1, *Tm*2, and *Tm*2*a* are effective[a]

Temperature (°C)	Heterozygous resistance	Homozygous resistance
17	55.8	87.7
22	37.8	79.0
26	19.3	72.0
30	14.1	61.4

[a] Calculated from the data of Cirulli and Ciccarese (1975). The data for the genes *Tm*1, *Tm*2, and *Tm*2*a* are averaged.

Table 3.2. Effect of increased temperature in reducing resistance to tobacco mosaic virus in tomato. The loss of resistance is partitioned between three temperature intervals, the entries being the change in the percentage of avirulent isolates over the temperature interval[a]

Temperature interval (°C)	Heterozygous			Homozygous		
	*Tm*1	*Tm*2	*Tm*2*a*	*Tm*1	*Tm*2	*Tm*2*a*
17–22	16	20	16	21	0	5
22–26	21	5	32	16	0	5
26–30	5	2	5	16	5	11

[a] Estimated from the data of Cirulli and Ciccarese (1975).

Table 3.3. Effect of temperature on the dominance of resistance in tomato to various tobacco mosaic virus isolates, dominant resistance being expressed as a percentage of total resistance[a]

Temperature (°C)	Resistance gene		
	*Tm*1	*Tm*2	*Tm*2*a*
17	53	80	58
22	45	56	44
26	14	50	12
30	0	34	7

[a] Calculated from the data of Cirulli and Ciccarese (1975), the entries in the table being the number of isolates avirulent on tomato plants heterozygous for resistance as a percentage of the number avirulent on plants homozygous for resistance.

22° C. The first entry in the table is thus 42–26 = 16%.). This confirms that point, already discussed in detail for wheat stem rust, that the transition temperature from effectiveness to ineffectiveness of a resistance gene is not directly related to the cardinal optimum temperature for the infection of susceptible varieties. The transition temperature is variable, determined by the hydrophobicity of the resistance protein of the host and of the virulence protein of the pathogen and reflecting, by hypothesis, the transition from dissociation to association of the relevant protein polymers.

Thermodynamically, mass action is as relevant as the dissociation constant of a protein polymer. Temperature affects the dissociation constant; homozygosity affects polymerization by mass action. The protein coded for by an allele for resistance copolymerizes less readily (i.e., is less hydrophobic) than that coded for by the allele for susceptibility; resistance reduces copolymerization. Homozygosity for resistance should therefore act in the same direction as reduced temperature, which also reduces copolymerization. It does in fact. This is brought out by Table 3.1.

Mass action, coupled with endothermic susceptibility, makes for a reversal of dominance; if there is a change of a gene's dominance it should be in the direction of dominant resistance at lower temperatures and recessive resistance at higher temperatures. This is what is in fact found. For example, at 17° C about half of the resistance conditioned by the gene *Tm* 1 was dominant and half recessive, while at 30° C none of the resistance was dominant, Table 3.3 gives further details.

3.6 The Temperature Effect in Some Other Diseases

As far as possible this chapter keeps to pathogens that have been demonstrated to relate to the host on a gene-for-gene basis. However, in order to cast the net a little wider, we include in this section, besides diseases named as gene-for-gene diseases in the literature, some examples of monogenic resistance that are probably within the orbit of our discussion although adequate proof is lacking.

Tobacco cultivars carrying the *N* gene from *Nicotiana glutinosa* are resistant to tobacco mosaic at 22° C, but susceptible at 28° C. The phenomenon was first observed in *N. glutinosa* itself by Samuel (1931), and found to be general in all tobacco species and cultivars carrying the *N* gene (McKinney and Clayton, 1945). One must see this against the background that in susceptible varieties, i.e., those without the *N* gene, tobacco mosaic virus RNA is much more infective and stable at 20° C than at 28° C (Kassanis and Bastow, 1971). Here again is evidence that the effectiveness or failure of the resistance gene in the host cannot be ascribed to the weakness or strength of the pathogen.

Pea *(Pisum sativum)* plants homozygous for the recessive resistance gene are resistant to bean yellow mosaic virus at both 18° C and 27° C; and those homozygous for susceptibility are susceptible at both these temperatures (Schroeder et al., 1960, 1966).

Heterozygous plants were mostly resistant at 18° C, but susceptible at 27° C. Here is both the expected temperature effect and the expected mass action effect expressed as a reversal of dominance, the resistance gene being dominant at 18° C, but recessive at 27° C. The results conform with those for tobacco mosaic virus, discussed in the previous section.

Bean *(Phaseolus vulgaris)* plants with monogenic resistance to common bean mosaic virus are resistant when the air temperature is 25° C by day and 29° C by night, but become susceptible at 32° C (Thomas, 1954).

Spinach *(Spinacia oleracea)* plants with monogenic dominant resistance are resistant to cucumber mosaic virus 1 at 16, 20, and 24° C, but not at 28° C (Pound and Cheo, 1952).

Cotton plants *(Gossypium hirsutum)* with the genes *b* and *BN* are resistant to *Xanthomonas malvacearum* at temperatures of 25.5° C by day and 19° C by night, but susceptible at the higher temperatures of 36.5° C by day and 19° C by night (Brinkerhoff and Presley, 1967). At still higher temperatures, 36.5° C by day and 26.5–29° C by night, the process reverses, a matter discussed in the next section.

Powdery mildew of wheat follows the standard pattern. A recessive gene in wheat varieties Hope and Renown (derived from H-44, closely related to Hope) is effective against appropriate races of *Erysiphe graminis tritici* at temperatures up to 20° C, but ineffective at 24° C, at which temperature a type "4" reaction is produced (Futrell and Dickson, 1954; McIntosh et al., 1967).

The dominant resistance gene *An* makes blue lupin *(Lupinus angustifolius)* plants resistant to *Glomerella cingulata* at temperatures up to 27° C, but the plants become moderately susceptible at 29° C, and fully susceptible at 32° C (Wells and Forbes, 1967). *G. cingulata* is considerably thermophilic, with optimal growth in vitro from 25 to 29° C, according to literature cited by Wells and Forbes. The plants with the *an* gene for susceptibility were mostly resistant at temperatures up to about 21° C, but were fully susceptible at 24–32° C, the highest temperature tested. The small difference between the turning-points, 21° C with the gene *an* and 27° C with the gene *An* suggests that the proteins specified by these two genes differ little in hydrophobicity.

Resistance in tomatoes to the root-knot nematode *Meloidogyne* resembles resistance to many virus, bacterial, and fungus pathogens, in that it is associated with hypersensitivity. The tomato cultivar Nematex with the dominant gene *Mi* for resistance is resistant to *M. incognita acrita*, *M. javanica*, and *M. arenaria thamesi* at 28° C but susceptible at 33° C (Dropkin, 1969). This confirmed an earlier report that resistant tomato cultivars, with unspecified resistance genes, were resistant at 20 and 25° C, but susceptible at 34.5° C (Holtzmann, 1965). This temperature, 34.5° C conditioning susceptibility is well above the optimum for attack; and at 34.5° C plants without the resistance gene were much less infected than at 25 and 30° C. The optimum temperature for infection of plants without the resistance gene was about 27° C. Here too we can reasonably infer than the failure of the resistance gene at 34.5° C was not the result of increased aggressiveness of the nematode.

3.7 Some Seemingly Anomolous Examples Considered

There is a minimum temperature below which disease does not occur, an optimum temperature most favorable to disease, and a maximum temperature above which disease does not occur. The response/temperature curve is ordinarily skewed to the right, the change from optimum to maximum temperature being more abrupt than the change from minimum to optimum.

The literature of the effect of temperature on stripe rust of wheat caused by *Puccinia striiformis* is contradictory. Gassner and Straib's (1934) experiments have been widely quoted as evidence that resistance in *P. striiformis* reacts differently to temperature from resistance to other cereal rusts. They tested wheat lines in greenhouses with mean temperatures of 14.8° C (with a mean range from 12.8 to 17.0° C), 21.0° C (with a mean range from 17.2 to 25.3° C), 23.0° C (with a mean range from 20.3 to 26.1° C), and 24.3° C (with a mean range from 20.0 to 27.2° C). They used three races of *P. striiformis*: 2, 7, and 9. German winter wheat varieties were mostly susceptible, with reaction types "3" and "4", at 14.8° C, mostly still susceptible at 21.0° C, but mostly resistant at 23.0° C and fully resistant at 24.3° C. German summer wheat varieties followed much the same pattern with race 9. With races 2 and 7 they were mostly susceptible at 14.8° C, mostly resistant at 21.0° C, and fully resistant at 23.0 and 24.3° C. Most South American wheat varieties were susceptible at 14.8 and 21.0° C, and some were still susceptible at 24.3° C, but mostly there was a swing to resistance at 23.0° C. The North American wheat varieties were susceptible at 14.8° C, and lost little of their susceptibility at 21.0 and 23.0° C; but at 24.3° C resistance was evident, especially to race 2.

All this on the surface seems evidence for a swing from susceptibility to resistance with rising temperatures, until one examines the results against the cardinal temperatures for infection by *P. striiformis*. Then one finds that what Gassner and Straib were calling resistance was in fact the reaction at or above the maximum temperature at which *P. striiformis* can infect.

Butler and Jones (1949) put the optimum temperature for *P. striiformis* at 11° C and this agrees well with the average temperatures during the rust period of countries where *P. striiformis* is prevalent. Newton and Johnson (1936) in Canada found that the optimum for germination of uredospores was 10–12° C, good germination occurring only between 5 and 18° C. The optimum for rust development was 13–16° C. Stroede (1933) found the optimum temperature for the germination of uredospores to be about 11° C. After 6 h almost 100% of the spores germinated at 11–12° C, 32% at 16.5–17.5° C, trace amounts at 19–20° C, and none at 23° C. Even after 24 h there was no germination at 23° C. At 25–26° C there was no germination even after 72 h.

Against this background we see that Gassner and Straib's lowest greenhouse temperature, 14.8° C, was already at or above the optimum for infection; and that temperatures at which he found susceptibility to change to resistance were at or above the cardinal maximum temperature. To think in terms of resistance of the host at temperatures so high as to inactivate the parasite is to carry belief in artifacts to the point of absurdity. It stretches the legitimate use of terms too far when the host plants above the maximum cardinal temperature for disease are called resistant. For this reason we reject Gassner and Straib's results as not being relevant to our discussion, or contrary to the trends established in the previous two sections.

There is nothing in the literature to suggest that *P. striiformis* behaves abnormally over a temperature range up to, say, 19° C, which could be considered relevant. *P. striiformis* is a low-temperature fungus, and its behavior must be judged accordingly.

Another anomalous example is bacterial blight of cotton plants *(Gossypium hirsutum)* caused by *Xanthomonas malvacearum*. Plants with the resistance genes *b* and *BN* are resistant at temperatures of 25.5° C by day and 19° C by night, but susceptible at the higher temperatures of 36.5° C by day and 19° C by night (Brinkerhoff and Presley, 1967). So far, this is in keeping with wheat stem rust and all the other diseases. At still higher temperatures, however, 36.5° C by day and 26.5–29° C by night, the process reverses, and plants become resistant again. These higher temperatures averaging (if averages are permissible) above 32° C compare with an optimum temperature for *X. malvacearum* of 25–30° C (Elliott, 1951); and doubt will remain about the significance of the reversal at high temperatures until it is determined how abruptly the aggressiveness of *X. malvacearum* falls off with temperatures above the optimum. Alternatively, the possibility of a reversal through monomer inactivation at high temperatures cannot be excluded. This was the topic of Section 3.3, and is raised again in the next section (Sect. 3.8).

The literature of leaf rust of wheat caused by *Puccinia recondita* is confused. Because of the work of Gassner and Straib (1932), leaf rust is commonly stated in the literature to change toward a susceptible reaction type at low temperatures. Further examination is needed.

Gassner and Straib (1932) claimed that the reactions of the varieties Malakoff and Norka to race 14 changed from infection type "0" at 18.7° C to types "3" and "4" at 6° C. Two comments are relevant.

First, the conditions were highly artificial. In particular, Malakoff and Norka had to be kept "more than one month" at the low temperature to give the susceptible reaction, whereas susceptible varieties (such as Kanred) gave the reaction within 3 weeks *at the same temperature*. One infers from this that Malakoff and Norka were still resistant after 3 weeks, the relevant incubation period, but collapsed later. There is a prima facie case here that the observed reactions of Malakoff and Norka were those of senescence or other response to artificial conditions and not those of low temperature. (One of the colored photographs Gassner and Straib published shows senescence clearly.) It is well known that in detached leaves of Khapli wheat senescence destroys resistance to stem rust; there is a large literature on this.

Second, Gassner and Straib's observations under artificial conditions are flatly contradicted by observations in the field. Chester (1946) records that in Kansas in 1928 flecks without uredinia survived the winter, to change to sporulating lesions in the spring. Again, in 1938, a leaf rust epidemic was preceded by the occurrence of great numbers of chlorotic flecks without uredinia. Later these changed to sporulating lesions [3]. Flecking is characteristic of resistance (type "0"), and here in Chester's observations are clear examples of resistance at low temperatures changing to susceptibility at higher temperatures.

There is other evidence of greater resistance in winter. Mains and Jackson (1926) stated that the variety Hussar was often highly resistant to some races when inoculated in the fall and winter, but only moderately or slightly resistant in late spring. They also noted that Kanred wheat and a number of durum and emmer varieties showed a "much higher" resistance in the field than they did in the glasshouse; and glasshouse temperatures were noted to be higher. Waterhouse (1929) observed that the varieties Carina and Hussar showed flecks in the winter months but infection type "4" in summer, when infected with Australian race 1. Other varieties to change from resistance in winter to susceptibility in summer were Hope, Iumillo, and Thew.

Turning now to results other than those involving low winter temperatures, we may condense the findings of Roberts (1936), Hassebrauk (1940), and Newton and Johnson (1941). Except for reactions on the varieties Brevit, Carina, and Hussar there were no large and consistant changes comparable with those mentioned in the previous two paragraphs. In Brevit, Carina, and Hussar the changes of reaction were in the direction of greater susceptibility at higher temperatures, thus conforming with the results recorded for other cereal rusts. In the other varieties studied, the changes were smaller and somewhat irregular. This is not just a subjective summary of a confused literature. Chester (1946) was able to write that many workers had observed that Brevit, Carina, and Hussar are consistently variable in their reactions, much more so than the other five differential varieties (Malakoff, Webster, Loros, Mediterranean, and Democrat), and several had expressed the opinion that Brevit, Carina, and Hussar were unsuitable as differentials because of this. Chester was able to draw up a table (his Table 8) greatly simplifying race groups by omitting reported reactions on Brevit, Carina, and Hussar.

[3] These seem to be the first recorded examples of the fact that in cereal rusts the mechanism of resistance is not one that destroys the pathogen. From lesions typical of resistance—flecks, etc.—the rust fungus can be made experimentally to grow into leaves of susceptible varieties and there to show susceptible reactions (Sharp and Emge, 1958; Burrows, 1960; Seevers et al., 1971; Littlefield, 1973).

Johnson and Schafer (1965) found that the varieties Colotana, Lageadinho, Frontana, and La Prevision 25 were more susceptible to races 1 and 2 of leaf rust at 27° C than at 16° C, with 21° C being intermediate in reaction. Rajaram et al. (1971) found that the variety W 3301 was resistant at 15–18° C but intermediate ("X" reaction) at 24–27° C. Jones and Deverall (1977) found that the gene *Lr* 20 was effective at 20.5° C, partially effective at 26° C, and ineffective at 30.5° C.

Rajaram et al., however, identified a second kind of resistance. The varieties W 3300 and W 3303 were susceptible at 15–18° C but resistant at 24–27° C. Chester (1946) records that there is a concensus among workers that leaf rust develops most rapidly in wheat fields when the mean temperature is in the range 15–22° C. That is, varieties W 3300 and W 3303 are susceptible varieties at the usual temperatures of rapid leaf rust development, but resistant at temperatures where leaf rust is no menace even in susceptible varieties. Whether the genes concerned are related to the conventional *Lr* genes is an open question, and there is yet no evidence that they belong to a gene-for-gene system. We take the matter up again in the next section.

To summarize the evidence about the two kinds of resistance to leaf rust, the varieties Brevit, Carina, Hussar, Colotana, Lageadinho, Frontana, La Prevision 25, W 3301 and, possibly, Kanred, Hope, Iumillo, and Thew, and varieties with *Lr* 20 have resistance that follows the predicted pattern and decreases at higher temperatures. The varieties W 3300 and W 3303 have another kind of resistance that becomes operative only at supraoptimal temperatures, reversing the general pattern.

In principle there is no reason against there being more than one sort of resistance. Nor are chlorosis and necrosis diagnostic of a particular sort. The type strain of tobacco mosaic virus, producing the typical systemic mosaic at 20° C, produces chlorotic lesions indicative of resistance at 35° C, at which temperature there is only one-tenth as much virus in the tissues as at 20° C (Kassanis and Bastow, 1971).

3.8 Temperature Reversal

In Section 3.3 it was noted that at temperatures above ordinary field or room temperature, the tertiary structure of protein monomers begins to break down, and as a result the monomers cease to be able to polymerize; with the protein flagellin the process begins at 28° C and depolymerization is complete at 50° C. In the previous section we noted that some varieties of cotton are resistant to *Xanthomonas malvacearum* at medium temperatures, become susceptible at higher temperatures, and finally become resistant again at still higher temperatures. We also noted that some varieties of wheat, susceptible at normal temperatures, become resistant to leaf rust at high temperatures. Are the processes connected? An answer would be premature; but two matters are worth raising, very briefly.

The first relates to the cardinal maximum temperature for disease. The temperature at which high temperature resistance is complete is of necessity the cardinal maximum temperature, because there can be no disease at still higher temperatures. To what extent then does the inactivation of protein monomers determine the cardinal maximum temperature for disease? Is the cardinal maximum temperature set by the most unstable protein monomer? Can plant pathology show the way here for plant physiology? I leave the questions open.

The other matter relates to the gene-for-gene hypothesis. Is the high temperature resistance on a gene-for-gene system? If it is, this would perhaps suggest that there may be *lr* genes which have no known *Lr* alleles, and which therefore can be detected only through monomer inactivation. (To demonstrate a gene-for-gene relation it would be necessary to have at least two resistance genes of the same sort, i.e., resistance genes effective at high but not at low temperatures, and at least two genes for virulence matching these particular resistance genes at appropriately high temperatures. The argument is fully set out in Section 3.26. Analyzing a mixture of high temperature resistance genes and ordinary *Lr* genes would give spurious results.)

The essential assumption, that globular proteins vary in internal stability, poses no difficulty.

Nothing further will be said in this book about the inactivation of protein monomers at high temperatures. We shall confine discussion to protein polymerization at ordinary field temperatures, which we take to mean temperatures at which disease would freely occur in susceptible varieties of the host, if enough inoculum was present and if dew, relative humidity, and all other pertinent conditions favored disease.

3.9 Conditional Responses to Temperature. Dominance of Resistance as a Test

If temperature brings about a change of resistance in gene-for-gene diseases, it will be in the direction of greater susceptibility with rising temperatures. The conditional *if* is relevant. Protein polymerization is always endothermic, and will proceed further as temperatures increase (until monomer inactivation interferes). There is, however, nothing in thermodynamic theory to say that an observed response will occur within the narrow range between the cardinal minimum temperature at the lower end of the range and the cardinal maximum temperature at the upper end, i.e., there is nothing to say that a temperature response will in fact be observed. The wheat stem rust resistance gene *Sr* 6 is unstable; it conditions resistance at low temperatures, but susceptibility at high temperatures. The gene *Sr* 11 is stable; so far as the evidence goes, it conditions resistance at all temperatures. We infer that the thermodynamic parameters are such as to give an observed response with the gene *Sr* 6 at temperatures within the range at which stem rust occurs. With *Sr* 11 there is no response within this range; there is no turning-point below the cardinal maximum temperature for stem rust infection, and this cardinal maximum temperature is the upper limit of relevance. Of necessity, the examples discussed in previous sections were examples of responses within relevant temperature ranges.

Changes from recessive to dominant resistance must be among the various possibilities for further research. Recessive resistance, although not the rule at ordinary temperatures, is not rare; Biffen's (1905) demonstration of Mendelian inheritance of resistance concerned recessive resistance in wheat to *Puccinia striiformis*. If there is a change (note again the conditional *if*) one would expect it to be in the direction of a change from recessive resistance at higher temperatures to dominant resistance at lower temperatures. For on our thermodynamic theory of resistance in gene-for-gene diseases, dominance or recessivity of resistance in the host, and dominance or recessivity of avirulence in the pathogen are mass action effects with the equilibrium constant determined by temperature.

Examples of a reversal of dominance by temperature, with resistance dominant at lower temperatures but recessive at higher temperatures, are the behavior of tomatoes to tobacco mosaic virus and of peas to bean yellow mosaic virus, discussed in Sections 3.5 and 3.6.

Recessivity in resistance bears on the protein polymerization hypothesis in another way. Recessive resistance should tend to condition intermediate reaction types. The correlation is very loose because the reaction type is influenced among other things by background genes and temperature. However, in the experiments of Knott and Srivastava (1977) the correlation appeared. These workers analyzed cultivars of common wheat showing good resistance to stem rust in the International Spring Wheat Rust Nurseries, and isolated six new unidentified recessive genes, all of which conditioned only a very moderate level of resistance, with infection types "2^+3". Plant breeders for obvious reasons try to avoid using only moderate levels of resistance, and recessive genes at normal temperatures may well be more common than the accumulated data suggest. Because both dominance and reaction type depend on temperature, tests to determine whether they are correlated should be carried out at the same temperature.

The relevance of intermediate reaction types to the protein polymerization hypothesis was discussed in a different context in Section 3.4.

3.10 Protein Polymerization and the Solvent

We have been considering hydrophobicity and protein polymerization in terms of protein dissolved in water; but plant proteins are dissolved in cell sap. There are solutes other than the proteins. They affect protein polymerization. In general, solvent molecules will promote polymerization if they have a greater energetic affinity for the polymer than for the monomer.

Glycerol and ethanol in the solvent increase protein polymerization (Stevens and Lauffer, 1965; Jaenicke and Lauffer, 1969; Oosawa and Asakura, 1975). The mechanism does not seem very clear. It may have something to do with glycerol and alcohol having larger molecular radii than water. If this is so, one would expect a still larger effect with mannitol and the sugars. (See note on p. 82.)

In what follows we are concerned not with glycerol $CH_2OH \cdot CHOH \cdot$-CH_2OH but with its homolog mannitol $CH_2OH \cdot (CHOH)_4 \cdot CH_2OH$ and sugars such as glucose $CH_2OH \cdot (CHOH)_4 \cdot CHO$.

Mannitol in hypertonic solution is used in the culturing of isolated protoplasts, and presumably affects protein polymerization within the isolated protoplasts. Consider the experiments of Otsuki et al. (1972). They used protoplasts isolated from the mesophyll tissue of tobacco plants with and without the *N* gene for resistance to tobacco mosaic virus. The concentration of mannitol in the culture solution was 0.7 M, which is strongly hypertonic. The *N* gene is very sensitive to temperature, giving effective resistance at 22° C, but not at 28° C (see Sect. 3.6). Culturing isolated protoplasts destroyed resistance. When the isolated protoplasts were infected with tobacco mosaic virus, the protoplasts from both the resistant and the susceptible varieties were indistinguishable at both temperatures with respect to the rate and extent of virus production. No sign of necrotic cell death (the symptom of resistance) was detected in the protoplasts, even when the *N* gene was present and the temperature appropriate for resistance. In the mannitol solution cells with the *N* gene became fully susceptible. Otsuki et al. deduced from their findings that the expression of the *N* gene requires the interaction of invaded cells with surrounding tissues. That is, they ascribed the result to the cells being isolated.

We deduce, quite differently, that the expression of the *N* gene was changed by the interaction of invaded cells with the surrounding culture medium. In detail, mannitol caused the relevant proteins to polymerize, thereby changing resistance to susceptibility. Alternatively, loss of water by osmosis concentrated the sugars and other solutes within the protoplasts, thereby promoting polymerization and susceptibility.

Can we reasonably ignore the deduction of Otsuki et al. that the expression of the *N* gene requires the interaction of invaded cells with surrounding tissues? Tomiyama et al. (1958) sliced potato tubers having an *R* gene into discs of different thickness, and inoculated the discs with an incompatible race of *Phytophthora infestans*. If the discs were thinner than 1 mm, their resistance to the incompatible race was seemingly lost when they were inoculated with a sufficiently dense suspension of zoospores; but thick discs retained their resistance. Here we have evidence of an interaction of invaded cells with surrounding tissues. At first sight this seems to support the deduction of Otsuki et al., but closer inspection removes the support. In the experiments of Tomiyama et al., the invaded cells always reacted hypersensitively, with necrosis, irrespective of whether the discs were thin or thick. Disc thickness was relevant only to the further spread of the fungus through the surrounding tissue, a matter irrelevant to isolated cells in liquid culture. Moreover, disc thickness was relevant only when the inoculum dose was massive enough to cause infection of most of the surface cells. With smaller doses of inoculum, causing infection only here and there, the *R* gene conditioned resistance even in thin discs. The thickness effect found by Tomiyama et al. required the interaction of infected cells, also, it would seem, a matter irrelevant to isolated cells in liquid culture.

Our deduction, that mannitol or osmosis promoted protein polymerization and hence susceptibility in the isolated protoplasts in culture, is important in two ways. First, if the deduction is correct, it is independent evidence for the protein-for-protein hypothesis. Secondly, it opens up new avenues for research into the molecular chemistry of gene-for-gene systems. The same conditioning *if* applies. If culturing in a hypertonic mannitol solution (or other relevant solution) brings about a change, it should be in the direction of gained susceptibility. It is not certain that there will be a change. That depends on all the parameters. Thus it is conceivably possible, though by no means certain, that resistance in the cultured protoplasts of Otsuki et al. would have been maintained, if the temperature had been substantially lower than 22° C, or if the concentration of mannitol had been substantially less than 0.7 M. Isolated cell protoplasts may yet prove to be a flexible research weapon, where they are applicable.

3.11 Osmotic Pressure Differences

Mature tissue of the wheat plant differs greatly from immature tissue both in osmotic pressure (and all that this involves) and susceptibility to rust diseases. There may well be a parallel here between the phenomenon, just discussed, of the loss of resistance against tobacco mosaic virus in isolated protoplasts in hypertonic solution. The suggestion is that in mature plants decreasing osmotic pressure, indicating a decreasing concentration of sugars and other solutes in the cell sap, increases resistance to *Puccinia* spp. through the depolymerization of protein.

Newton and Brown (1934) expressed sap from fully developed leaves, still photosynthetically active, and from young leaves still folded within the uppermost sheaths of wheat plants. The osmotic pressure, determined from the vapor pressure, of the sap expressed from the young leaves was substantially greater than that of sap expressed from the fully mature leaves. For example, the osmotic pressure of sap from young leaves of the wheat varieties Acme, Iumillo, and H-44-24 averaged 11.2 atm while the corresponding figure for sap from mature leaves was 6.7 atm. They related this to the resistance which develops in wheat, oats, and barley to particular races of *Puccinia graminis* as the plants become older.

Johnson and Johnson (1934) compared the sugar concentration in young and mature leaf tissue. They defined the young tissue as comprising the young leaves still enfolded by the uppermost sheath and the young stem below the uppermost node, and mature tissue as the fully developed upper leaves and their adherent sheaths. The major difference was in reducing sugars (see Table 3.4), which on an average were 4.2 times greater in concentration in young than in mature tissue. Sucrose was also more abundant in young tissue, but the difference between young and mature tissue was less.

Although there was a great difference in sugar content between young, susceptible tissue and mature, resistant tissue there was no more sugar in susceptible varieties than in resistant varieties. Consequently Johnson and Johnson concluded that higher sugar concentration had no direct effect in promoting susceptibility. Lyles et al. (1959) confirmed this. Because there is no evidence for a direct relation between sugar concentration and resistance or susceptibility, the matter was allowed to drop. We are, however, interested in sugar (and other solute) concentration in wheat cell sap, not in relation to the carbohydrate metabolism of *Puccinia* spp., but in relation to protein polymerization. This necessitates a re-examination.

Table 3.4. Reducing sugar in young and mature tissue of wheat varieties, expressed as a percentage of fresh weight[a]

	Young tissue			Mature tissue		
	mean	highest	lowest	mean	highest	lowest
Resistant varieties[b]	1.84	2.58	1.11	0.48	0.82	0.25
Susceptible varieties[c]	2.20	2.62	1.58	0.48	0.52	0.45
Average	2.02			0.48		

[a] Calculated from the data of Johnson and Johnson (1934).
[b] Varieties exhibiting mature-plant resistance to prevalent races of *Puccinia graminis tritici*: Hope, H–44–24, Acme, and Pentad.
[c] Varieties susceptible at all stages to prevalent races: Marquis, Garnet, and Little Club.

Table 3.5. Reaction types of two varieties of wheat to two races of *Puccinia recondita* at two stages of growth[a]

Variety	Seedlings		Mature plants	
	Race 9	Race 161	Race 9	Race 161
Thatcher	4	4	0; 1	4
Red Bobs	4	4	4	4

[a] Data of Bartos et al. (1969). Reaction type 4 means fully susceptible, and type 0; 1 highly resistant.

The name, mature-plant resistance or adult-plant resistance, is usually applied to situations in which the seedling is susceptible and the adult plant resistant. The example to be discussed is one of vertical resistance specific to some races of *P. recondita*, but not others. Thus, the variety Thatcher has mature-plant resistance to race 9 of *P. recondita*, but not to race 161 (Bartos et al., 1969). Table 3.5 illustrates how the varieties Thatcher and Red Bobs are both susceptible as seedlings to race 9; Thatcher plants become resistant when adult, but Red Bobs plants remain susceptible. To race 161 both varieties are susceptible, both as seedlings and as adult plants.

Bartos et al. studied the genetics of virulence on adult plants of Thatcher in an F 2 population of cultures from a cross between races 9 and 161 (made on *Thalictrum speciosissimum*, the host of the sexual stage). A gene-for-gene relation was demonstrated, the resistance gene being recessive.

Earlier work by Newton and Brown (1934) demonstrated large differences in reaction type between young and mature tissue in the same plant. They used *P. graminis tritici*, *P. graminis avenae*, and *P. coronata avenae*. Many varieties of wheat and oats that were resistant throughout life were resistant only because old resistant tissue enfolded and protected young susceptible tissue from inoculation by spores. When spores were deposited in direct contact with young, rapidly growing tissue, susceptible ("3" or "4") types of infection were commonly found

Table 3.6. Average crown rust reaction types of two varieties of oats to two temperatures at four stages or growth[a]

Stage of growth	Var. Appler		Var. Mo. 0–205	
	15° C	25° C	15° C	25° C
Seedling	3.8	4.0	3.0	4.0
Juvenile	2.9	3.0	0.9	3.1
Boot	1.1	3.2	0	1.7
Anthesis	2.0	3.2	0	1.0

[a] Data of Simons (1954) for race 205. Reaction type 0=some necrosis or chlorosis but no uredia; 1=necrotic or chlorotic areas, a few of which contain small uredia; 2=necrotic or chlorotic areas, most of which have small or medium sized uredia; 3=chlorotic areas surrounding abundant medium sized uredia, no necrosis; 4=no chlorosis or necrosis, the uredia being large and abundant.

even in wheat and oat varieties normally considered resistant to the particular isolates of the pathogen. Resistance that increases with the age of the tissue or of the plant is common in cereal rusts. Even common wheat varieties like Marquis and Garnet (listed as susceptible in Table 3.4) have mature-plant resistance to appropriate races of *P. graminis tritici.* Adult Marquis wheat is resistant to race 38, and intermediate (with an "X" reaction) to race 120. Both these races give a susceptible ("3"–"4") reaction in young tissue, represented in Newton and Brown's experiments by the basal portion of the flag leaf and the lower portion of the sheaf of the flag leaf.

The balance between protein polymerization and susceptibility, on the one hand, and dissociation and resistance, on the other, seems to be rather fine, as one would expect if the balance is tipped by changes in the solvent. This could explain why recessive resistance seems to occur more commonly in mature-plant resistance than in many other gene-for-gene relations. The mature-plant resistance gene in Thatcher to race 9 of *P. recondita* is, it will be remembered, recessive. So too the gene *Lr* 12 in Exchange conditioning mature-plant resistance to race 5 of leaf rust is only partially dominant, and the gene *Lr* 13 is partially dominant in the variety Frontana, and recessive in the variety Manitou (Dyck et al., 1966).

Temperature affects mature-plant resistance as it does seedling resistance: higher temperatures, if they have an effect, increase susceptibility. The data of Simons (1954) on crown rust of oats illustrate this. The varieties Appler and Mo. 0–205 became more resistant with age to race 205 of *Puccinia coronata*, but at all ages plants at 25° C were more susceptible than at 15° C. The variety Appler has mature-plant resistance at 15° C but not at 25° C. Details are given in Table 3.6.

To return to the cell sap as a solution, the solutes can have two distinct roles in protein polymerization. Sugars and other solutes can, we assume, affect polymerization rates (as glycerol and ethanol do) and the dissociation constant of the protein copolymer. An alternative role could exist for amino acids. An abundance would, one expects, increase the rate of synthesis of host protein and thus, by mass action, decrease the amount of free pathogen protein and hence make for better compatibility between host and pathogen. There is little point in further discussion until the thermodynamic effect of cell sap on protein copolymerization has been studied except to predict that mature-plant resistance will turn out to be a complicated chemical phenomenon.

3.12 Analytical Difficulties Arising from Procedures

Evidence for polymerization or depolymerization of a protein extracted from a plant is not good evidence for polymerization or depolymerization in the plant before extraction. The solvents differ; and extraction changes the relevant environment. The presence of a polymer in an extract is no proof that it will be present in the living cell, nor is the absence of the polymer in the extract proof of its absence in vivo.

From what has been said of temperature effects, it is clear that the protein polymerization associated with susceptibility in gene-for-gene diseases is loose, the polymers having relatively very high dissociation constants. It is no guide to think in terms of protein polymers such as hemoglobin which have very small dissociation constants, resulting from tight noncovalent binding across relatively large surface areas of contact between sub-units. Hemoglobin and its like can be studied in vitro, and for this very reason are the sort of protein polymer that has been chosen for study extensively. Our problem is different. Isolation of the protein to determine the state of polymerization may, by the very fact of isolation, defeat its own purpose by changing the state of polymerization.

We are as much concerned with the state of polymerization as with the identity of the protein monomers concerned: with whether, say, 10% of the protein is in polymeric form, or 90%. Traditionally, plant pathologists have been differently involved. Evidence that a substance conditions resistance usually requires that the substance be isolated and shown to be antifungal or bactericidal. One famous research institution can boast with very justifiable pride that it detects, isolates, and determines the structure of antifungal compounds in plants at the average rate of once every two weeks. The problems of this chapter are different. Their elucidation will have to depend more on thermodynamics than on organic chemistry; and the evidence that in gene-for-gene diseases susceptibility is endothermic is a foretaste of the sort of research that is needed.

3.13 Isozymes and Electrophoresis

In plant pathology isozymes have almost invariably been taken to mean electrophoretically determined isozymes. Many isozymes cannot in fact be distinguished by current techniques of electrophoresis.

Changed electrophoretic mobility reflects a change in the net charge of the protein molecule; and a changed amino acid can be detected electrophoretically when the new amino acid carries a charge different from the one it replaces. What is the probability that the change will be detected? Of the 20 standard amino acids, 15 are neutral in charge, three have a positive net charge, and two have a negative net charge (Shaw, 1965). On the assumption that all the amino acids are equally abundant in the protein, it has been calculated that only about one-fourth (27.6%) of single substitutions in the nucleotides will produce a change in net charge (Shaw, 1965). Fitch (1966), looking at the matter somewhat differently, gave a slightly more generous estimate: of the possible missense mutations, 40% affect the net molecular charge of the monomer.

These figures show only part of the weakness of electrophoresis as a tool for research in gene-for-gene disease. Changes in the hydrophobic side chains are the least amenable to electrophoretic analysis, and it is precisely around these side chains that the hydrophobic effect centers. The fingerprinting that we regard as the basis of gene-for-gene recognition may well be outside the range of electrophoretic analysis. The change from susceptibility to resistance in the host (or from virulence to avirulence in the pathogen) probably has a larger effect than a change of fingerprints, but even so it seems possible to have a large change in the hydrophobic effect without a change in net electric charge. One cannot confidently expect an electrophoretic clue to changes in susceptibility in the host or in virulence in the pathogen.

Changes in size and configuration of the protein, as well as the net electric charge, affect electrophoretic mobility. For monomers these changes are likely to be small. If one uses the data of Chothia and Janin (1975) for the subunits of insulin, trypsin and horse oxyhemoglobin, one calculates that, on an average, an amino acid residue occupies less than 1% of the total surface area of the protein. The effect of a change here on mobility is likely to be small.

Polymerization brings about a great change in electrophoretic mobility; but here we are confronted with the difficulties discussed in the previous section. With finely balanced polymerization, one cannot assume that polymerization in vivo would run parallel with polymerization in the type of solution and at the range of concentration used in the technique of analysis by electrophoresis.

What have been determined electrophoretically as isozymes may well be clusters of isozymes, differing in hydrophobicity, but similar in net electric charge. For example, Daly et al. (1971) found the same electrophoretic isoperoxidase patterns associated with wheat stem rust resistance gene *Sr* 11 as with *Sr* 6. There is nothing in this finding to preclude a difference in hydrophobicity (which there probably is, to judge by the temperature stability of resistance conditioned by *Sr* 11 and the instability of resistance conditioned by *Sr* 6) or in fingerprinting (which there certainly is). The conclusion of Daly et al. that the isoperoxidases associated with *Sr*6 and *Sr*11 are identical is unwarranted, because it concerns only that relatively small fraction of isoperoxidases which can be detected by electrophoresis.

3.14 A Hypothesis About Susceptibility and Resistance

In susceptibility (our hypothesis runs) the pathogen excretes a protein into the host cell which copolymerizes with a complementary host protein. This copolymerization interferes with the autoregulation of the host gene that codes for the protein, and by so doing turns the gene on to produce more protein. This host protein in fungus disease serves primarily as food for the pathogen; it is the essence of parasitism that the host feeds the parasite. The copolymerization has a second function: it binds the protein entering the host from the pathogen, and takes it out of circulation as a catalyst.

In resistance (the hypothesis runs) the protein specified by the gene for avirulence in the pathogen and excreted into the host does not polymerize. It stays actively catalytic in the host cell as a foreign body, introducing catalytic processes not normal to the healthy host cell. It is the "elicitor" or "toxin" that catalyzes the start of the reactions that backlash and inactivate the pathogen, directly or indirectly, as by starvation.

The heading to this section says hypothesis, not hypotheses. There is a single hypothesis that, appropriately adapted, covers both susceptibility and resistance. There is a dualism. The same protein from the pathogen, isozyme variations apart, can, if copolymerized, make for susceptibility, and turn on a supply of food for the pathogen, or, if unpolymerized, act to promote resistance reactions. This dualism traces back to the relevant protein from the pathogen having at least two involved areas on its molecular surface. The one area becomes involved in copolymerization during susceptibility; the other area, in the absence of copolymerization, binds with the substrate to start the catalysis that eventually leads to resistance.

In catering for both susceptibility and resistance, the hypothesis has a great advantage over most hypotheses that see in susceptibility nothing more than nonresistance.

3.15 Feeding the Pathogen. Function of Genes for Avirulence

Consider, say, leaf rust of wheat, in a fully compatible combination giving a type "4" reaction. Observe two points: (1) spores are produced in mass, often over many days or weeks. These spores are rich in protein, and the ultimate source of this protein is the host plant. We infer, from massive spore production, that the pathogen is a heavy feeder. (2) with *Puccinia recondita*, there is remarkably little visible damage to host cells for many days before the final collapse. The pathogen establishes itself, sends some mycelium running into the neighboring tissue, and settles down to producing spores. It seems that the pathogen has managed to extract protein from the host cells without seriously damaging the normal processes that keep the cells alive, i.e., without seriously interfering with the proteins in the chromatin, in the general body structure, in the enzymes of respiration, and every other essential business of life. We can infer, too, that the pathogen has not injected into the host cells a mass of proteolytic enzymes that destroy indiscriminately (see Chap. 10). Indiscriminate digestion may, or may not, be a feature of maceration diseases such as soft rots; it is certainly not a feature of gene-for-gene diseases.

The question then is, how does a highly compatible pathogen extract protein in quantity, without greatly disturbing the host cells that are the givers of this protein? Presumably, it uses a source that in uninfected tissue is untapped or only very lightly tapped. It turns on genes that in healthy, undamaged tissue are at least partially repressed or regulated, and consumes the extra protein coded for by these genes after they are turned on.

There is a considerable literature of enzymes, such as peroxidase, produced in greater quantities as a result of infection by fungi and viruses. There is also a considerable literature suggesting what these enzymes do; peroxidase, for example, has been implicated in lignin formation and various other roles. The one suggestion that seems to have been omitted is that the enzymes are plain, protein food for the pathogen. Most of us are not specially concerned about the identity of the proteins in a T-bone steak we eat. Why then assume that the pathogen cannot cope with a particular enzyme, not as enzyme but as protein food?

There are probably at least 20 loci in wheat involved in susceptibility/resistance to stem rust. They are the sites of the *sr* or *Sr* genes. We write "at least", because new genes are being discovered year by year as the search for resistance becomes wider. Also, there are possibly many loci with *sr* genes that have never mutated to *Sr* genes, and have therefore remained undetected and, on present techniques, undetectable. For ease of presentation, accept the round figure of 20 genes and leave it at that.

During infection (our hypothesis runs) these 20 genes are turned on—derepressed—to produce protein. It is the matching protein of the pathogen that does the turning on, by interfering with the process of autoregulation (Sect. 3.17). Earlier (in Sect. 2.7) it was argued, mainly on grounds of population genetics, that the genes for avirulence in the pathogen are not suicide genes, but have some function essential to the pathogen. Here we suggest what this essential function is: the genes are needed to extract protein from the host plant. We see too that the number of matching loci involved in a gene-for-gene system (20 in our provisional figure for wheat stem rust) is determined by what genes the pathogen, not the host, has available. The more genes the pathogen can use for this purpose, the more efficiently it can extract protein.

Genetical opinion (Price, 1976) has already been quoted (in Sect. 2.6) to the effect that repetition (duplication) of genes coding for proteins might apply only for genes of a few proteins needed in abundance at critical times. This meets our case precisely. The critical time for the host is when the pathogen invades it, and the proteins needed in abundance at this time are crisis proteins not involved in normal metabolism. Their removal from the host cell would therefore meet the requirement that in compatible combinations, as in wheat stem rust or wheat leaf rust, protein removal by the pathogen should cause minimal damage to normal host metabolism. As to the repetition of genes, the facts seem reasonably clear. In wheat, for example, in addition to an estimate (for convenience) of 20 genes matched gene-for-gene by *Puccinia graminis*, we must add possibly another 20 matched independently by *P. recondita*, eight by *P. striiformis*, ten by *Tilletia caries*, *T. foetida*, and *T. contraversa*, six by *Erysiphe graminis*, and six by *Mayetiola destructor*, the hessian fly. The exact numbers cannot now be determined. Further work may show that some of the matching by the different pathogens is not independent; but even so the evidence suggests that there is ample repetition.

Pathogens need much else besides protein. They need minerals, carbohydrates, and other foods, a matter clearly established for cereal rusts by Mains (1917). However, these other foods do not enter into our discussion; they have no direct role in determining gene-for-gene relations, although they may have an indirect role through protein polymerization (see Sect. 3.10). Free amino acids have also been ignored for various reasons, among others because the increased RNA, swollen nucleoli and other apparatus of protein synthesis in infected cells suggest that the traffic is primarily in proteins.

3.16 In Fungus Disease, Is Wound Protein the Host Protein Involved? The Function of Genes for Susceptibility

Are genes for susceptibility/resistance those that code for wound proteins? The question must be asked, although the answer at present is of secondary importance. Let us get the issues clear. Disprove that the process governing susceptibility is endothermic, and our hypothesis as a generalization collapses. Disprove that the protein involved is a wound protein, and all that is needed is to substitute another protein. The protein's identity is a detail that can be adjusted if further research requires it.

Wound proteins are suggested because they are proteins needed for defense and repair, which are not tasks normal to healthy unwounded tissue; and they are needed in a hurry. All this argues the need for the plant cells to have a large enzyme-producing capacity not normally used, but available in sudden emergencies. Considerable repetition of normally repressed genes answers this need. This fits the evidence that in gene-for-gene diseases there is considerable repetition of genes for susceptibility/resistance.

It was argued in Section 2.5, mainly on evidence from population genetics, that the gene for susceptibility in the host is not primarily a gene of invitation to the parasite—it is not a gene for self-inflicted harm—but has an essential useful function in the host. The evidence, we now suggest, is that this useful function is defence by means of enzymes against the consequences of wounding, the defense including repair of damage.

The pathogen, it seems, turns the tables on the host. The enzymes the host makes for its defense, the pathogen uses for its protein food.

Various proteins accumulate at or near wounds. Peroxidase has received the most attention. The literature has been reviewed by Lipetz (1970) and Uritani (1971). Peroxidase seems to be synthesized de novo from amino acids (Kawashima and Uritani, 1963; Kawashima et al., 1964; Kanazawa et al., 1965). It is the enzyme that has received most attention in wounding, but some caution is needed in interpreting this: Lipetz (1970) suggests that some of peroxidase's current popularity as a research topic might perhaps be ascribed to the relative ease with which the enzyme can be determined by analysis. Polyphenol oxidase (Kosuge,

1969) and phenylalanine ammonia-lyase (Minamikawa and Uritani, 1965) are other proteins that increase at cut surfaces and in infected tissues.

There is a substantial literature of increased peroxidase content as a result of infection by fungi in gene-for-gene diseases (for example, Macko et al., 1968; Seevers and Daly, 1970; Daly et al., 1971; Seevers et al., 1971, for stem rust of wheat; Hislop and Stahmann, 1971, for powdery mildew of barley). Some of this literature will be discussed later, in different contexts. There is also a substantial literature of peroxidase, especially preinfectional peroxidase, in relation to horizontal resistance, and a literature of peroxidase in relation to diseases not known to be gene-for-gene diseases. This literature is not relevant to our present topic.

We shall have occasion to use peroxidases as a vehicle for discussion, but it does not seem to be of the greatest importance to identify the protein involved. Indeed, there is no reason to believe that only one kind of protein is involved in a particular host–pathogen combination, or that the same kind of protein should necessarily be involved in different combinations. What seems important in wheat leaf rust or other highly compatible combinations (we are discussing susceptibility!) is that the pathogen should be able to remove large amounts of protein from the host cell without greatly disturbing the normal functioning of the cell. Proteins that are extra to normal requirements, as wound proteins must be, seem to be the most probable candidates for involvement.

It might be objected that the release of wound protein into the host cell would disturb the normal functioning of the cell by starting wound-repairing reactions. The answer is that copolymerization with pathogen protein would destroy the ability of the host protein to catalyze. The protein would no longer be peroxidase (or polyphenol oxidase, or other wound enzyme), but would be a catalytically inert copolymer which could be taken through the host cell safely. Copolymerization, as we see it, would do two things: it would cause the host to produce protein by interfering with autoregulation (the topic of the next section), and it would make that protein safe to transport by making it no longer a catalyst.

3.17 Autoregulation of Gene Expression

The autoregulation of gene expression is one of the central problems of genetics. In eukaryotes especially, relatively few of an organism's genes are active at any given time. As the organism develops, genes are turned on or off, as required and at the proper time. How is this done?

The most widely discussed model is the operator–repressor model developed for prokaryotes. This requires two genes, the extra gene being required to produce a cytoplasmic repressor. Evidence for the operator–repressor model is well known for some systems, but there is also evidence for systems in which there is no second gene involved in regulation. The gene product itself participates directly as a cytoplasmic factor in controlling the expression of its own gene. Systems of this sort occur in eukaryotes. The need for only one gene makes them consistent with gene-for-gene systems.

The autoregulation of gene expression has recently been discussed by Goldberger (1974) and Calhoun and Hatfield (1975). In an autoregulated system the regulatory gene product controls its own synthesis, and may or may not play other metabolic roles within the cell. The allosteric first enzymes interact specifically with the transfer RNA, and there are various suggestions about how the first enzyme-transfer RNA complexes could regulate gene expression. Details need not concern us. What does concern us is that the autoregulated system is essentially a system of feedback inhibition. As the first enzyme is removed, by the copolymerization which is the essence of our hypothesis, more of this enzyme would be produced simply because it is removed as fast as it is produced. Parasitism and copolymerization, we hypothesize, remove feedback inhibition and turn on a supply of host protein which would otherwise not be turned on.

The genes coding for peroxidase are evidently regulated, because in healthy, unwounded tissue the peroxidase content is relatively low; the regulation appears to be autoregulation, because gene-for-gene systems are most easily explained if the host genes act individually; and, on the evidence that susceptibility is determined endothermically and that protein–protein association is one of the very few chemical ways of storing and releasing information about interactions involving large numbers of host and pathogen genes, protein copolymerization seems to be the likely method of freeing the production of peroxidase from autoregulation. The arguments hold not just for peroxidase, but for any other proteins that increase markedly in amount as a result of wounding or infection.

3.18 A Corollary

The autoregulation hypothesis has a corollary: the host protein that copolymerizes during compatible infection need not exist in the host in detectable amounts before infection. According to the autoregulation hypothesis, it is infection that turns the genes for susceptibility/resistance on; before infection they are repressed. Their coded proteins may therefore not exist in the healthy plant, or may exist only in trace amounts.

Nor is it inevitable that the coded host protein as part of a copolymer would occur in substantial quantity after infection. If, as we suggest, the coded host protein is consumed as food by the fungus, its accumulation would represent the balance between production and consumption; and that balance might not be large.

To turn to the parasite, one might assume that its production of the relevant enzyme would begin only when production is needed. The enzyme need not necessarily exist in detectable amounts in ungerminated spores or mycelium not in immediate contact with the parasitized protoplasts.

Autoregulation could be the reason why conventional analytical procedures, requiring the isolation and identification of relevant substances, have not so far been successful in explaining compatibility in gene-for-gene diseases, except, possibly, for brome mosaic virus in barley (Sect. 3.24) and examples of common antigenic surfaces in host and pathogen (Ch. 4).

3.19 Comparison with Antigen–Antibody Systems

In Section 2.14 it was suggested that the antibody system in mammals and the protein-for-protein system in gene-for-gene plant disease involve the same method of protein recognition. Consider the matter further. The plant pathogen's "antigen" is the relevant protein it excretes, to stimulate the host's "antibodies" by (we have just suggested) interfering with autoregulation. In animals the antibody suppresses the pathogen. In plants (our hypotheses runs) the pathogen consumes the "antibody". The stories are in essence the same; just the endings differ. They differ, perhaps, largely because animals have a blood stream, and can mobilize antibodies from afar, tipping the balance in favor of the antibodies.

3.20 Peroxidase Production in Infected Susceptible Plants

There seems to be no way of measuring how much peroxidase is produced. In susceptible host plants the increase of peroxidase after infection measures production less consumption; and because there is no way at present of measuring consumption, there is no way of measuring production. Nor can we safely arrive at an estimate indirectly by studying peroxidase production in resistant host varieties after inoculation. Admittedly, in resistant varieties in which the pathogen is inactivated there is presumably little or no consumption, so that peroxidase content probably follows production more closely. But here another variable is introduced; in resistant varieties, host cells are more greatly injured, and injury alone can stimulate peroxidase production. Moreover, the peroxidase stimulated in resistant varieties may be chemically distinct. Seevers et al. (1971) found the electrophoretically detected "isozyme 9" to increase markedly when wheat with resistance gene *Sr* 6 or *Sr* 11 was inoculated with avirulent race 56 of *Puccinia graminis tritici* at 20° C, but not when wheat with the corresponding susceptibility genes *sr* 6 and *sr* 11 was inoculated.

As one might expect from the concept that peroxidase is possibly food for the pathogen, there is with wheat stem rust a preliminary increase of peroxidase concentration in susceptible host varieties after inoculation, followed by a slowing of the rate of increase, or even by a decrease, once sporulation begins (Seevers and Daly, 1970; Daly et al., 1971; Seevers et al., 1971), and consumption is high.

3.21 The Lag Phase in the Resistance Reaction

For the purpose of illustration, suppose as before that there are 20 loci involved in wheat stem rust, i.e., 20 *sr* or *Sr* genes in the host and correspondingly 20 avirulence or virulence genes in the pathogen. In susceptibility, all 20 *sr* genes in both host and pathogen are matched and compatible. If there are *Sr* genes, they are matched by corresponding virulence genes. After compatible infection the pathogen excretes its 20 isozymes, each with its own fingerprints, and by copolymerization these turn on the production of protein coded for by the *sr*/*Sr* genes. This is the story so far for compatibility.

Suppose now that at one of the 20 loci there is incompatibility: An avirulence gene in the pathogen is matched with a resistance gene in the host. Their proteins do not polymerize. There is no longer compatibility at all 20 loci. There is compatibility at 19/20 and incompatibility at 1/20 of the loci; 19/20 of the host protein–pathogen protein pairs copolymerize, and 1/20 does not.

Compatibility processes dominate for a while after inoculation, and this shows as a lag in the onset of recognizable resistance reactions. At 20° C in wheat with the gene *Sr*6 inoculated with race 56 of *Puccinia graminis tritici*, which is avirulent, it takes about 4 days for flecks to appear (Seevers and Daly, 1970), flecking being the first macroscopic sign of host cell inactivation, about 5 days for the respiration rate in resistant wheat to differ much from that in susceptible wheat (Antonelli and Daly, 1966), and about 3 days for a change of decarboxylation activity by indoleacetic acid to occur (Antonelli and Daly, 1966). Time lags of varying length have been widely reported in the literature of fungus disease.

A time lag occurs in bacterial disease as well. Klement et al. (1964) infiltrated tobacco leaves with incompatible species of pathogenic *Pseudomonas (P. syringae, P. mori, P. phaseolicola)*. The lag was sufficient for the bacteria to multiply about 100-fold before incompatibility processes intervened and stopped further multiplication. Some more recent literature has been reviewed by Lyon and Wood (1976). These results (we note parenthetically) suggest that protein-for-protein relationships may be far wider than the relationships detectable by genetic analysis of gene-for-gene systems.

Equally important was Klement et al.'s finding that with purely saprophytic pseudomonads there was no lag. The bacteria failed to increase at all; they were neutral. Presumably the enzymes excreted for a purely saprophytic existence do not include a protein needed to turn the host DNA on.

There is evidence that *Erwinia* spp. (Lakso and Starr, 1970) and *Pseudomonas* spp. (Lelliott et al., 1966), which macerate host tissue and cause soft rots, also fail to multiply and induce hypersensitivity. With macerating bacteria, too, the excreted enzymes would not be expected to include a protein to turn the DNA on; and in any case macerating bacteria tend to destroy the host cell's nucleus as soon as, or before, they destroy the host's cell wall. Destruction of the nucleus means that there would be no organized DNA left to turn on.

3.22 Uniformity in Resistance Reaction Types. Secondary Effects

Incompatibility responses are surprisingly uniform. In the cereal rusts, e.g., the same system of reaction types ("0", "1", and "2" for resistance) is used for stem rust of wheat and oats, for leaf rust of wheat, and crown rust of oats, among others.

This uniformity suggests that the basic processes of resistance in these rust diseases are uniform. Whereas most of this chapter has been devoted to differences in response, we must now examine mechanisms that make for similarity.

Consider incompatibility. A gene for avirulence in the pathogen is matched by a gene for resistance in the host. The temperature responses suggest that, with incompatibility, relevant copolymerization does not occur. The protein specified by the gene for avirulence in the pathogen enters the host, fails to copolymerize, and remains as a foreign, active enzyme in the host cell. It is a catalyst. Now for the first time in this chapter we must consider isozymes of the pathogen not as carriers of genetic variation manifested, through copolymerization, in a large number of pathogenic races, but as uniformly acting catalysts. This is inherent in the definition of isozymes, as genetically controlled variants sharing a common catalytic process. Uniformity through catalysis is just as intrinsic a manifestation of isozymes as is the variability that gives isozymes their name.

To explain the uniformity of resistance reactions we look therefore to catalysis, not polymerization. On the protein-for-protein hypothesis, the enzyme specified by the gene for avirulence and let loose by the pathogen through the membranes stays loose and unpolymerized (or incompletely polymerized) in the host cell, and as a catalyst promotes the reactions that lead to incompatibility and resistance, and promotes them with considerable uniformity. There is a very large literature of these reactions; but the incompatibility reactions, although unquestionably important, are secondary. Incompatibility results in various manifestations which may include the death of the host cell, and the formation of lignin, phytoalexins, and other products often associated with hypersensitivity. Phytoalexins are discussed briefly in a later chapter. What concerns us here is that in gene-for-gene diseases the primary decision whether host and pathogen are compatible or not is taken, according to our hypothesis, at protein level; and that if the decision is for incompatibility, some uniformity in secondary reactions is enforced by uniform catalysis.

There is some contrary experimental evidence that requires scrutiny. Doke and Tomiyama (1975) concluded that the hypersensitive death of potato cells caused by infection with an incompatible race of *Phytophthora infestans* does not depend on dynamic protein synthesis. They used an inhibitor of protein synthesis, blasticidin S, to inhibit protein synthesis in the cut surface of potato leaf petioles. Their analyses, involving the incorporation of ^{3}H-leucine, showed that protein synthesis was indeed inhibited in the petiole tissue. Blasticidin S was applied to the cut petiole surface before it was inoculated with an incompatible race of *P. infestans*. This treatment had no effect on the hypersensitivity reaction following infection; hence they concluded that the reaction does not depend on protein synthesis.

Against this evidence is the fact that blasticidin S is a very selective antibiotic inactive against many fungi, including *P. infestans*. One assumes that in the treated and inoculated petioles the fungus continued to synthesize protein normally. Therefore (for those who favor this sort of evidence) Doke and Tomiyama's results, far from being contrary evidence, support our hypothesis that resistance results from unpolymerized, or incompletely polymerized, pathogen protein in the infected host cell. Further, the experiments of Tani et al. (1976), who also used blasticidin S, indicated that, contrary to Doke and Tomiyama's results with potato blight, de novo protein synthesis was required for the expression of resistance in oats to *Puccinia coronata avenae*.

Enzymes are ubiquitous. Even the processes that lead from DNA to coded first protein are catalyzed in sequence by enzymes. Evidence that proteins, whether produced de novo or otherwise, are not needed in resistance reactions is therefore likely to be treated by many with reservation.

3.23 Temperature Limits Again. The Genes *Sr*6 and *Sr*11. Effect of Ethylene

It can be accepted as self-evident that changes in resistance can be seen only at temperatures between the cardinal minimum temperature for the disease and the cardinal maximum (see Sect. 3.2). If the thermodynamic parameters require a marked change of hydrophobic effect either below the minimum or above the maximum, the change is biologically imaginary.

It can be accepted as probable that changes in resistance, between the minimum and maximum temperature, can occur anywhere within this range. Some experimental evidence for this was given for tobacco mosaic virus in tomato in Section 3.5, Table 3.2. More evidence about this would be welcome, because it goes to the core of protein polymerization. Unfortunately much evidence has probably been missed, through lack of appropriate screening. Screening, of necessity, has been tied largely to the needs of practical plant breeding, and concentrated on the temperature range at which epidemics are likely to occur. Thus, with wheat stem rust, e.g., a wheat variety that changed from resistant to susceptible to a relevant race of the pathogen at 5–7° C would likely be recorded as susceptible without further question, and a variety that changed from resistant to susceptible at 32° C would likely be recorded as resistant without further question, because workers usually concentrate on the range 15–25° C in their studies of susceptibility and resistance. For a detailed probe of the thermodynamics of resistance a wide range of temperature needs further study, and especially the lower end of the range at which results are unlikely to be complicated by monomer inactivation (Sect. 3.3).

The wheat stem rust resistance gene *Sr*6 is sensitive to temperature, losing its effectiveness at high temperatures. The gene *Sr*11 is insensitive. Daly et al. (1971) have argued that the two genes must therefore be essentially different, biochemically as well as genetically. We see the matter differently, for reasons already given: with the *Sr*6 system, the relevant copolymerization occurs below the cardinal maximum temperature for disease; with the *Sr*11 system, it does not. The matter, as we see it, is one of thermodynamic parameters, tracing back to the amino acid residues at the polymerizing surface. At any given temperature the association constant in the reversible monomer/polymer process is greater in the *Sr*6 system than in the *Sr*11 system[4].

Daly et al. (1971) found that treating wheat with ethylene had a differential effect in the *Sr*6 and *Sr*11 systems. At 20° C ethylene (80 ppm) causes *Sr*6 to become ineffective as a resistance gene, while *Sr* 11 remains effective. Relevant details are given in Table 3.7. In this, too, Daly et al. see an essential difference between the *Sr* 6 and *Sr* 11 systems. Again, we see it much more simply. If ethylene stimulates relevant protein production in the host, then, by mass action, more of the relevant protein derived from the pathogen would be copolymerized when the association constant was large than when it was small. Put differently, ethylene is more likely to reduce the amount of protein derived from the pathogen and existing free and unpolymerized in the host, if the association constant is high; and (from what was said in the previous paragraph) there is reason to believe that the association constant is higher in the *Sr* 6 than in the *Sr* 11 system.

If what is said in the previous section is correct, there should be an ethylene/temperature interaction, with the possibility that *Sr* 6 types of wheat would remain resistant to avirulent races of the stem rust fungus in the presence of ethylene, provided that the temperature was lowered sufficiently. An ethylene/temperature interaction, if it occurs, would be experimentally important, in that it would increase the scope of studies on temperature effects and hence our understanding of the thermodynamic processes.

Table 3.7. The effect of ethylene on the stem rust resistance of near-isogenic lines of wheat prossessing either *Sr* 6 or *Sr* 11 alleles[a]

Allele	No ethylene	Ethylene 80 ppm
sr 6	S[b]	S
Sr 6	R	S
sr 11	S	S
Sr 11	R	R

[a] Data of Daly et al. (1971). The temperature was 20° C, and the fungus race 56.
[b] S = susceptible; R = resistant.

3.24 In Virus Disease, Is RNA Replicase the Protein Involved?

In virus disease the evidence implicating RNA replicase as the protein involved in copolymerization is much stronger than that implicating peroxidase in fungus disease. The clues lead straight to RNA replicase.

[4] There is a possible alternative explanation. It is possible that in the *Sr* 6 system the polymerizing surface has a higher proportion of amino acid residues with aliphatic side chains, because Kauzmann (1959) has shown that the effect of temperature is greater with these residues than with the phenylalanine residue.

In Sections 3.5 and 3.6 evidence was led to show that gene-for-gene virus disease conforms with other diseases in the essential tests: if temperature affects resistance, it affects it in the direction of a change from resistance to susceptibility at higher temperatures. If temperature affects dominance, it affects it in the direction of a change from dominant resistance to recessive resistance at higher temperatures. The viruses concerned, it will be remembered, were tobacco mosaic virus in tobacco plants and in tomato plants, bean yellow mosaic virus in pea plants, common bean mosaic virus in bean plants, and cucumber mosaic virus 1 in spinach plants; and results consistently conformed with expectation.

The similarity of response to temperature in fungus, bacterial and virus diseases must not be taken to mean that the same proteins are involved. It means that susceptibility is endothermic and, probably, that the same process, copolymerization, is involved; but it does not necessarily indicate more than that.

In tobacco mosaic virus, less than 10% of the RNA is accounted for by the coat cistron. After tobacco has been infected, two proteins, which are not coat protein, appear early in the infected protoplasts. The evidence is that they are RNA replicases (Hunter et al., 1976). Their molecular weights are 165000 and 140000. The sum of these molecular weights exceeds the coding capacity of the virion RNA. Hunter et al. suggest that there must be overlapping somewhere. A simpler explanation would be that the proteins are copolymers, of which part is coded by the virus genome and part by the host genome. This explanation is consistent with circumstantial evidence about brome mosaic virus in barley plants and direct evidence about the $Q\beta$ phage in *Escherichia coli*.

Hariharasubramanian et al. (1973) found by means of gel electrophoresis a protein, apparently the product of the viral genome, in barley plants infected with brome mosaic virus. The protein, which was not a coat protein, was synthesized at an early stage after infection, and was associated with cell fractions rich in brome mosaic virus RNA replicase activity. They suggested that this protein might be part of the larger RNA replicase molecule characterized by Hadidi and Fraenkel-Conrat (1973), i.e., that the brome mosaic virus RNA replicase is a copolymer, part of which is specified by the genome of the host and part by the genome of the virus.

Kondo et al. (1970) and Kamen (1970) worked with a bacterial virus: the $Q\beta$ phage of *Escherichia coli*. The $Q\beta$ replicase is a tetramer of which one monomer is coded by the virus, and three by the host genome. Because the replicase has been purified and studied in vitro, some of its properties have been determined. What is specially important is that the polymerization properties seem to tally precisely with what has been predicted in our protein–protein hypothesis. The dissociation constant is high, i.e., there is a loose association–dissociation equilibrium between the replicase and its subunits. Kamen studied sedimentation, and found that the four subunits co-sedimented in large aggregates in the absence of ammonium sulfate in the solvent, but separated in high salt concentration (0.2 M). Kondo et al. observed that the four subunits occur in nonstoichiometric and variable amounts within the sharply sedimenting enzyme band. These in vitro studies do not indicate exactly the state of aggregation in vivo, because of differences in the solvent; but they do indicate a looseness of binding.

The evidence is that what we call the gene—the RNA cistron—for avirulence/virulence in the virus should be identified as the gene coding for a subunit of the virus RNA replicase. This coding is the avirulence gene's useful and necessary role in the pathogen which in Section 2.6 was deduced on grounds of population genetics to exist.

A need to contribute only part of the replicase puts at the disposal of the virus a greater mass of enzyme than the whole of the virus genome could code for on its own. Mass for mass, a virus can be a highly efficient thief of host protein, even though there may be no more room in the genome than can accomodate a single gene for avirulence/virulence.

Competitive catalysis between virus and host RNA replicases would divert synthesis of host RNA to synthesis of virus RNA, which could explain why Reddi (1963) found that tobacco mosaic virus RNA was accumulated at the same time as host ribosomal RNA was depleted. In turn, this could explain why host ribosomes are degraded in infected tobacco plants. One could, however, debate about causes and effects: Reddi suggests that the degradation products of host RNA are used in the synthesis of virus RNA; it is equally possible that efficient enzymatic synthesis of virus RNA robs the host of its RNA, and that the symptoms of the disease follow from this robbery.

3.25 Summary and Conclusions

The foundation of parasitism is that the parasite obtains food and other requirements from the host. The question is, how does it do it? In this and the previous chapter we attempt to answer this, but only in relation to diseases which have been demonstrated to be gene-for-gene diseases or (to widen the scope) which are not improbably gene-for-gene diseases. Further, we are concerned primarily with host–pathogen recognition, and confine attention to those aspects of parasitism which, on available evidence, bear on recognition.

A hypothesis is discussed which in its simplest form equates host–pathogen compatibility, i.e., susceptibility, with protein copolymerization. Host protein copolymerizes with pathogen protein. There is available a substantial body of knowledge about the thermodynamics of protein polymerization. Protein polymerization is known to be an endothermic process, except at temperatures high enough to initiate the denaturation of proteins. The protein copolymerization hypothesis of host–pathogen compatibility can be tested on the basis that susceptibility ought to be an endothermic process. If temperature changes resistance, the change ought to be in the direction of resistance at lower temperatures and susceptibility at higher temperatures, and the turning-point ought not to be related to the optimum temperature for disease in susceptible varieties of the host plant. If temperature changes dominance, this ought to be in the direction of dominant resistance at lower temperatures and recessive resistant at higher temperatures.

The protein copolymerization hypothesis passes these tests; and the bulk of evidence is substantial enough for the hypothesis to be used in further discussion.

On the protein copolymerization hypothesis, in its form of a protein-for-protein hypothesis, Flor's gene-for-gene hypothesis becomes a matter of protein-for-protein recognition. On chemical theory currently accepted, recognition requires that for polymerization the hydrophobic effect must be strong enough to bring the constituent monomers together, the surfaces in contact must be contoured so that they fit closely together, and the N, O, and S atoms at the contact surfaces must be properly positioned to make the necessary hydrogen bonds. Presumably the problem of protein–protein recognition will be the subject of further research by chemists, and current theory may be extended or modified; but that does not concern us deeply at present, so long as we know that there is a recognition system.

Protein–protein recognition is the foundation of most antigen–antibody systems; and it was pointed out earlier that the protein-for-protein hypothesis uses nothing with regard to protein copolymerization and recognition that is not nowadays taken for granted in the antigen–antibody concept. We are on safe ground here.

A point mutation from avirulence to virulence in the pathogen implies a change to an amino acid with a more hydrophobic side chain, and one from susceptibility to resistance in the host implies a change to an amino acid with a less hydrophobic side chain.

In eukaryotes especially, gene activity is regulated so that no more of a product is made than is needed at any particular time. Autoregulation is a known process; the gene's product regulates the gene's activity, turning the gene on when more of the product is needed, and off when there is enough. There are possibly several different systems of autoregulation. What matters here is that in any system in which the first protein regulates, or is part of a complex that regulates, the gene's activity, the act of removing the protein must of itself turn the gene on. Binding the host's protein with the pathogen's protein in a copolymer is a way in which the pathogen can remove the host's protein, turn the gene on, and get still more of the protein. (The alternative method of removing host protein would be for the pathogen to destroy the host's protein with proteolytic or other protein-degrading enzymes. This would cause havoc in the cell, and destroy the whole purpose of the exercise. There is really no apparent alternative to copolymerization.)

The pathogen's own protein turns on the supply of host protein. This is why genes for avirulence/virulence have an essential role in the pathogen, even as the avirulence allele (Sect. 2.7).

An essential condition for the production of relevant protein by the host when autoregulation is disturbed is that in susceptible (compatible) host plants the nucleus must survive for a period after the cell is infected. (We are for the moment discussing nothing but compatibility.) This condition was studied in Section 2.4; it is met. It seems probable, in some diseases at least, that cytoplasmic protein coded for in organelles is also involved; but if the nucleus dies, the cytoplasm dies, so it is enough to concentrate attention on the nucleus.

Up to this point we have discussed matters thought to be general for gene-for-gene diseases; but in many ways these diseases are a heterogeneous group not amenable to generalization.

At one extreme we may consider leaf rust of wheat. As Allen (1926) pointed out, the mycelium of *Puccinia recondita* does not spread far into host tissue after infection; the pustule remains localized. *P. recondita* sits at table and lets its food be brought to it. The bringing is very efficient, and *P. recondita* is able to produce masses of spores from supplies brought in from the outside of the pustule. It was suggested that peroxidase might be the protein involved (or one of the proteins involved because there is no a priori reason why only one sort of protein should be collected by copolymerization). A reason for suggesting peroxidase is that it seems to have a part in the host plant's response to injury or wounding; if this is so, the host could probably make it available quickly and in quantity, and transport it. This would allow the pathogen to use the host's transport mechanisms.

Similar to *P. recondita* are other cereal rust fungi. Ward (1890, 1902, 1904) and Mains (1917) had a clear picture of parasitism in cereal rusts. Mains worked with *P. sorghi* in maize and *P. coronata* in oats. He noted that, although most cells of the leaf may be invaded by large haustoria of *P. sorghi*, yet no harmful effect is shown by the host until after some period of time. The fungus sends its mycelium into the intercellular spaces and then its haustoria into adjacent cells; nevertheless the invaded cells retain the characteristics of cells of uninfected tissues. Mains' special interest was carbohydrate metabolism, and he reached the conclusion expressed for *P. recondita* in the previous paragraph: the rust fungus lets its food be brought to it. He commented that the rust fungus, instead of attacking and killing the cells of the tissue in which it is situated, has a very different effect on them. While it is withdrawing food, at the same time it stimulates the infected tissue so that this loss of food is in turn compensated by the withdrawal of food from neighboring uninfected tissue.

It is because these fungi, initially at any rate, do not noticeably injure the cells they invade that we conclude that they use proteins coded by genes not normally turned on in healthy cells, or turned on only slightly. The fungus feeds without any apparent effect, initially, on the protoplasm, which indicates that the normal protoplasm of a healthy cell is not its food. In turn, this leads straight to autoregulation.

To interpolate a comment here, it is remarkable how up to date is the work of Allen, Mains, and Ward, among others. For an integrated anatomical picture of cereal rust infection their work, with ordinary light microscopy, has never been bettered, and seldom equalled by moderns with their electron microscopes. Anatomical cereal rust research has followed a course of learning more and more about less and less, coupled, often, with forgetfulness about the great body of knowledge available more than half a century ago.

Potato blight has an expanding lesion. *Phytophthora infestans* must fetch its own food. The host nucleus in infected cells of susceptible hosts does not long survive the advent of sporulation, and *Phytophthora infestans* is a less specialized parasite than *Puccinia recondita*.

In systemic plant disease a virus does not require a transport system for what it takes from the host, except on an intracellular scale. It was suggested that the virus takes from the host, by copolymerization, subunits of the virus RNA replicase, and that the gene for avirulence/virulence in the virus coded for the remaining subunits. The path to identifying the virus RNA replicase as the protein involved was via tobacco mosaic virus, brome mosaic virus in barley, and the Qβ phage.

There is much to indicate some profitable directions for future research. The body of evidence about the effect of temperature on resistance and dominance in gene-for-gene diseases needs to be extended. Temperature as an independent variable in controlled light intensity has relatively few complications, although it is not entirely free from them (e.g., low temperatures, 7° C or less, may cause the hydrolysis of starch and so alter the cell sap, perhaps relevantly to protein polymerization). The essential difficulty is that, thermodynamically, the range of temperature is great but, biologically, experiments are necessarily limited to temperatures between the cardinal minimum and maximum for disease. There is no inherent thermodynamic reason why relevant changes, from resistance to susceptibility or from dominance of resistance to recessivity, should occur within these limits. (Hence the reason for repeatedly stressing the conditional *if*: If there is a change,...) However, relatively so little research has been carried out on temperature effects on resistance and dominance of resistance that further research should give almost undiminished returns.

Susceptibility and resistance should both respond to temperature; and decreased temperature ought to change susceptibility to resistance just as increased temperature changes resistance to susceptibility, if there is a change. On first thoughts, this might seem to be a way of widening the field of temperature research; but second thoughts reveal a fundamental difficulty. Susceptible varieties have their full complement of *lr*, *sr*, and other alleles for susceptibility. If one were to study the effect of decreasing the temperature, one would not be able to distinguish between a turning-point from susceptibility to resistance, which involves protein depolymerization, and a cardinal minimum temperature, for which (on current scanty knowledge) no depolymerization is implied. Either way, the turning-point would be the minimum temperature. It is only when, as with *Glomerella cingulata* in *Lupinus angustifolius* (Sect. 3.6), the turning-point is far higher than the cardinal minimum temperature for growth of the pathogen in culture, that we suspect that the minimum temperature for disease coincides with protein depolymerization.

It is a tenet of science that one should develop the simplest and most economical hypothesis that will explain the known facts. In terms of our topic, we need two hypotheses, one for susceptibility and one for resistance of the host; and the two hypotheses should be interrelated, because susceptibility and resistance are interrelated.

The hypothesis of susceptibility propounded here is very economical. Protein copolymerization means susceptibility. On this basis, we explain how the fungus feeds itself or the virus acquires enzyme sub-units; and the hypothesis fits the facts of temperature effects on resistance and dominance of resistance. It also uses the elementary detail of molecular biology that mutations are efficiently stored chemically in sequences of synthesis up to the primary coded proteins.

The hypothesis of resistance propounded here is equally economical. Failure of the pathogen's protein to copolymerize leaves a foreign enzyme free in the host cell to start catalytic processes that end in incompatibility. Catalysis by enzyme molecules as distinct from the polymerization of enzyme molecules destroys most of the chemically stored information on mutation, and this could explain why resistance responses in gene-for-gene diseases are generally uniform.

Nevertheless, these hypotheses seem inadequate in two respects. They take no account of quaternary structure, which is at present unknown. They leave the host's allele for resistance with a purely negative role and (if we assume autoregulation) repressed. Only from one aspect does this negative role seem plausible. Resistance in gene-for-gene disease is exothermic. (In a reversible process, if susceptibility is endothermic, resistance is automatically exothermic.) This means that the resistance reaction in gene-for-gene diseases does not absorb energy; and this could possibly tally with a negative role for the reaction.

The fact that phage Qβ RNA replicase is a tetramer which comes three parts coded by host and one part by the phage reminds us that nothing in the gene-for-gene or the protein-for-protein hypothesis requires equal contributions of protein from host and pathogen. The reference to protein-for-protein is to the kinds of protein, and not to the number of monomers involved in polymerization. It is not implied in the gene-for-gene hypothesis that the gene in the host and the gene in the pathogen are turned on to produce protein at equal rates. About rates there is no evidence at all, one way or the other.

Because a molecular one-for-oneness is not implied, it might perhaps give less scope for misinterpretation if the hypothesis were referred to as the protein copolymerization hypothesis instead of the protein-for-protein hypothesis. The headings of Chapters 2 and 3 use the latter alternative, but only in order to align the hypothesis with the gene-for-gene hypothesis, and not to suggest a preference.

3.26 Identifying Gene-for-Gene Disease

Diseases are being recorded in the literature as gene-for-gene diseases on information that is inadequate and unacceptable. We link the protein-for-protein hypothesis with the gene-for-gene hypothesis; and it is absolutely essential that we know what diseases we are talking about. Therefore we conclude the chapter with a statement of the minimum requirements for the identification of a disease as a gene-for-gene disease.

To establish a gene-for-gene relation there must be virulence genes in at least two loci in the pathogen matching at least two resistance genes in the host. Nothing less will do. By Flor's enunciation (Sect. 2.3) of the gene-for-gene hypothesis, for every gene determining resistance in the host there is a specific and related gene determining pathogenicity in the pathogen. This is the essence of the gene-for-gene hypothesis; and the necessary evidence cannot be obtained unless there are at least two loci for resistance in the host and two for corresponding virulence in the pathogen.

Table 3.8 considers the hypothetical example of two resistance genes $R1$ and $R2$ at loci 1 and 2 in the host, and two virulence genes $p1$ and $p2$ at loci 1 and 2 in the pathogen. Race 1 of the pathogen, with $p1$, can attack $R1$-types of the host, but not $R2$-types; and race 2, with $p2$, can attack $R2$-types but not $R1$-types.

Table 3.8. Reaction of resistance genes at two loci in the host with corresponding virulence genes at two loci in the pathogen in a Flor gene-for-gene system. Hypothetical data

Virulence gene	Resistance gene	
	R 1	*R* 2
*p*1 (in race 1)	Susceptible	Resistant
*p*2 (in race 2)	Resistant	Susceptible

There is a differential interaction, and Table 3.8 gives the absolute minimum evidence needed to establish that for every gene determining resistance in the host there is a specific and related gene determining pathogenicity in the pathogen.

To establish that the relation in Table 3.8 is indeed the minimum, we shall use as an example the leaf blight of maize caused by *Helminthosporium turcicum*. The example is chosen because it is topical: *Helminthosporium* leaf blight has been passed as a gene-for-gene disease, on insufficient evidence, in a top-ranking journal, and illustrates the inadequacy of the "quadratic check" as a test.

Hooker (1963) isolated a type of resistance conditioned by a single dominant gene *Ht* 1. This gene has been tested in many countries, and has remained effective in all of them except the Hawaiian Islands. Here an isolate of *H. turcicum* was obtained that could overcome the resistance of gene *Ht* 1. Since then another resistance gene *Ht* 2 has been discovered that conditions resistance to all known populations of *H. turcicum*, including that from Hawaii. Lim et al. (1974), using techniques for producing perithecia and isolating ascospore progeny, demonstrated that virulence in *H. turcicum* to the monogenically resistant *Ht* 1-type of corn is conditioned by a single gene in the fungus. They concluded that the host–pathogen system, *Zea mays–H. turcicum*, fits the gene-for-gene model of Flor. Their conclusion, be it noted, is based on a single locus for resistance/susceptibility in the host: the *Ht* 1 locus. The *Ht* 2 locus falls outside the scope of their analysis, as it must because no virulence on it is known. Also, their conclusion is based on virulence at a single locus in the pathogen, for the same reason that virulence at a second locus, to match *Ht* 2, is unknown.

The evidence for their conclusion, that the host–pathogen system is a gene-for-gene system, is inadequate. To illustrate this, consider a counter suggestion, made only for argument's sake. This counter suggestion is that both resistance and virulence are quantitative, not qualitative as demanded by a gene-for-gene hypothesis. On this counter suggestion, the resistance given by gene *Ht* 2 is of the same kind as that given by gene *Ht* 1, but there is more of it. So too, on the pathogen's side, the pathogenicity of the Hawaiian isolate is (on this counter suggestion) of the same kind as that of the common isolates, but there is more of it. The resistance given by *Ht* 1 is enough to counter the pathogenicity of all isolates other than the Hawaiian isolate. The extra resistance given by the gene *Ht* 2 is, according to the counter suggestion, enough to overcome all known pathogenicity, including the extra pathogenicity of the Hawaiian isolate. This counter suggestion fits every detail of known fact; it fits known fact at least as well

Table 3.9. Reaction of resistance genes *Ht* 1 and *Ht* 2 in maize with a common isolate and a Hawaiian isolate of *Helminthosporium turcicum*[a], with hypothetical additions

(A) Data Showing Qualitative Resistance

Isolate	Resistance genes	
	Ht 1	*Ht* 2
Common	Resistant	Resistant
Hawaiian	Susceptible	Resistant
Undiscovered, hypothetical	Resistant	Susceptible

(B) Data Showing Quantitative Resistance

Isolate	Resistance genes	
	Ht 1	*Ht* 2
Common	Resistant	Resistant
Hawaiian	Susceptible	Resistant
Undiscovered, hypothetical	Susceptible	Susceptible

[a] From data of Lim et al. (1974).

Table 3.10. Reaction of an allele for susceptibility *r* 1 and an allele for resistance *R* 1 at a locus in the host with an allele for avirulence P 1 and an allele for virulence *p* 1 at a locus in the pathogen. The quadratic check

Pathogen allele	Host	Allele
	r 1	*R* 1
P 1	Susceptible	Resistant
p 1	Susceptible	Susceptible

as Lim et al.'s suggestion that resistance and pathogenicity are qualitative traits. The data of Lim et al., which are all the data available at present, fail to distinguish between qualitative and quantitative resistance or pathogenicity; and this failure traces back to there being only one relevant locus in the pathogen for virulence, in the present instance on *Ht* 1.

To illustrate these matters, suppose that a new isolate of *H. turcicum* were obtained which is virulent on the gene *Ht* 2. The question is, would it be virulent on the gene *Ht* 1 or not? The two possibilities are set out in Table 3.9. In the upper half of the table, resistance is qualitative, and the last two lines of part A of Table 3.9 match Table 3.8 in detail. *Ht* 1-types are susceptible to the Hawaiian isolate, but resistant to the new hypothetical isolate; and *Ht* 2-types are susceptible to the new hypothetical isolate, but resistant to the Hawaiian isolate. For each of the two genes determining resistance in the host there is a specific and related virulence in the pathogen; and this is the barest minimum evidence of a gene-for-gene relation. In the lower part of Table 3.9 (part B) resistance is quantitative. The new hypothetical isolate is virulent on both *Ht*1 and *Ht*2. The virulence is not specific and related, and therefore does not conform with a gene-for-gene relation. The discovery of a new isolate conforming with Table 3.9 (part B) would destroy Lim et al.'s contention of a gene-for-gene relation.

The confusion has arisen from what is called the quadratic check. The check is illustrated in Table 3.10. It involves only one locus in the host and one in the pathogen, and therefore contradicts our minimum requirement of two loci in both host and pathogen if a gene-for-gene relation is to be established. Yet in the literature the quadratic check is taken to be a check on whether gene-for-gene conditions are met. This cannot be accepted for two reasons.

First, the quadratic check cannot distinguish qualitative from quantitative resistance in the host, or qualitative from quantitative pathogenicity. For example, the check would be compatible with quantitative relations if, in a toxin-producing pathogen, pathogenicity meant more and nonpathogenicity less of the same toxin; and resistance meant more and susceptibility less tolerance by the host of the toxin. The same point has already been made about the quadratic check by Day (1974, his Fig. 4.1).

Second, trying to identify a gene-for-gene relation by the quadratic check is an exercise in bad logic. A gene-for-gene relation means that a quadratic check can be devised. But the converse is not necessarily true. That a quadratic check can be devised does not necessarily mean that there is a gene-for-gene relation. (A tree is a plant, but a plant is not necessarily a tree.)

Leaf blight of maize caused by *Helminthosporium turcicum* is not the only disease that has been wrongly classified, in the sense that the evidence is inadequate, as a gene-for-gene disease because of the quadratic check. Leaf blight of oats caused by *H. victoriae* is another; and there is an important group of plant diseases, discussed in Chapter 8, for which there is inadequate evidence to show whether resistance is qualitative or quantitative, vertical or horizontal.

Note Added in Proof. A reference has come to light that substantiates the suggestion on p. 57. Shelanski, Gaskin, and Cantor [Proc. Nat. Acad. Sci. U.S.A. **70**, 765–768 (1973)] have shown that sucrose and glycerol strongly promote the polymerization of protein, at concentrations relevant to our topic.

Chapter 4 Common Antigenic Surfaces in Host and Pathogen

4.1 Introduction

The previous two chapters have dealt with the hypothesis that in gene-for-gene systems there are complementary surfaces on the proteins of the host and of the pathogen, and that, given the necessary hydrophobicity, these surfaces associate for the proteins to copolymerize. The present chapter deals with common (shared) antigenic surfaces in host and pathogen, that is, with surfaces which are similar or identical. On the face of it, there is a sharp contradiction here. Similar surfaces, as distinct from complementary surfaces, do not associate; and common antigens in host and pathogen would, if commonness as such is stressed, clash with the hypothesis of copolymerization.

Fortunately, the contradiction is superficial, and tenable only if we assume a single associating surface on the host protein, and a single associating surface on the pathogen protein, so that the greatest polymer would be a dimer. With at least two associating surfaces on the proteins from both host and pathogen, so that trimers or higher polymers could be formed, complementarity and similarity could coexist; common antigenic surfaces would then become little more than a corollary of the protein copolymerization hypothesis. This chapter should thus be seen as a direct continuation of the previous two chapters.

In this reference to the number of associating surfaces we do not count surfaces in contact with a substrate for binding and catalysis. Catalysis does not enter this chapter; and associating surfaces are taken to mean surfaces that associate in polymers or copolymers.

To begin with, the chapter considers only gene-for-gene diseases or diseases possibly on a gene-for-gene system. Later the scope is widened.

4.2 Common Antigens in the Flax–Flax Rust System and Some Other Systems

Doubly et al. (1960) studied globular proteins from flax *Linum usitatissimum* and the flax rust fungus *Melampsora lini*. They extracted protein fractions from healthy plants of various varieties of flax, and from the uredospores of various isolates of the fungus. They prepared antisera in rabbits. When host and pathogen were compatible, with the flax variety susceptible to the rust isolate, the titers of rust antiserum against flax antigens were relatively high (1:160 or 1:320). When host and pathogen were incompatible, with the flax variety resistant to the rust isolate, the titers were relatively low (1:20 or 1:40). Susceptibility reflected a substantial quantity of antigens common to host and pathogen, and the introduction of a resistance allele reduced the quantity. These results were confirmed by Peterman (1967).

It is clear from the parallel results that the antigenic surfaces in the experiments of Doubly et al. are essentially the same as those discussed in the protein copolymerization hypothesis; and the hypothesis must accommodate the fact of common antigenic surfaces.

Similar serological relationships were reported for cotton *(Gossypium hirsutum)* and *Xanthomonas malvacearum* by Fedotova (1948), who used extracts from seeds, and by Schnathorst and De Vay (1963) and De Vay et al. (1967), who used extracts from leaves. Wimalajeewa and De Vay (1971) compared the antigens of *Ustilago maydis* with those of maize *(Zea)* and barley *(Hordeum)*. *U. maydis* and maize, which is susceptible, had antigens in common; *U. maydis* and barley, which is resistant, had no detectable antigens in common. Oats *(Avena)* were intermediate. A diploid line of *U. maydis* attacked 3-day-old oat seedlings, but not 6-week-old plants; and, conforming with this, the diploid line of *U. maydis* shared antigens with the 3-day-old seedlings, but not with the 6-week-old plants. De Vay et al. (1967) found antigenic determinants common to sweet potatoes *(Ipomoea batatas)* and *Ceratocystis fimbriata*; and Golik et al. (1977) found antigenic determinants common to potatoes *(Solanum tuberosum)* and *Synchytrium endobioticum*. McClure et al. (1973) showed that the serological relationships hold also for animal parasites of plants. Antisera for eggs of the nematode *Meloidogyne incognita* cross-reacted with antigen from cotton *(Gossypium)* and soybean *(Glycine)* roots; and antisera for cotton and soybean roots cross-reacted with antigens from eggs and larvae of *M. incognita*. It will be remembered (from Sect. 3.6) that susceptibility of tobacco to *Meloidogyne* is endothermic.

4.3 Coexistence of Complementary and Similar Surfaces

The essence of the protein copolymerization hypothesis is that the associating surfaces pack closely together and make hydrogen bonds between N, O, and S atoms (Sect. 2.14). The surfaces complement each other. The essence of the common antigen story is that the surfaces are similar or identical. Similar surfaces are unlikely to complement each other. Two convex surfaces, for example, could not pack closely, except at their tips. Hydrogen bonds between N, O, and S atoms would require the corresponding surfaces to have something like mirror imagery for the reactive atoms; and surfaces that are mirror images are unlikely to be similar.

Suppose that the monomer (*a*) from the host plant has a single associating surface *a* with the necessary hydrophobicity, that the monomer (*b*) from the pathogen also has a single associating surface *b*, and that the surfaces *a* and *b* are complementary. The two monomers could then form the dimer (*a*)(*b*). Note two matters. First, a dimer is the largest polymer possible of (*a*) and (*b*). Second, the complementary surfaces *a* and *b* are unlikely to be similar; common antigens are unlikely to occur.

Now suppose that the monomers from host and pathogen each have two associating surfaces *a* and *b*, with complementarity between *a* and *b*. Association could make a dimer (*ab*)(*ab*), a trimer (*ab*)(*ab*)(*ab*), a tetramer (*ab*)(*ab*)(*ab*)(*ab*), and so on, the number of units being determined by the free energy. (For simplicity, head-to-tail polymers have been used for illustration, but are not an essential part of the argument. Other geometrical patterns can occur when monomers have two or more associating surfaces.) Note here that the surfaces are both complementary, *a* to *b*, and similar, *a* to *a* and *b* to *b*. The copolymerization hypothesis and the existence of common antigens are reconciled.

Is this reconciliation realistic? There is direct evidence both for the frequency of polymerization beyond dimerization, and for the coexistence of complementary and similar surfaces. Three examples suffice. Urease, the first enzyme to be crystallized and shown in 1926 to be a protein, is a hexamer with a molecular mass of 480000. The monomers of the enzyme, lactate dehydrogenase, are of two sorts, A and B, with separate gene loci. The tetrameric enzyme exists in all five possible combinations, A_4, A_3B, A_2B_2, AB_3, and B_4, which neatly illustrates the coexistence of complementarity and similarity. Coexistence of the same sort is known in hemoglobin. In very young human embryos the hemoglobin tetramer is either $\alpha_2\varepsilon_2$ or ε_4 or, later, $\alpha_2\gamma_2$ or γ_4, where the Greek letters represent monomers of different sorts from different gene loci.

Common antigens, it seems, are an integral part of the protein copolymerization hypothesis, with the proviso that the monomers must have two or more associating surfaces.

4.4 The Intrinsic Experimental Difficulty

Our knowledge of common antigenic determinants is at present confined, for obvious experimental reasons, to preformed antigens, whereas we really need to know about antigens formed after infection, because these are the antigens concerned in copolymerization.

Consider the flax-flax rust system. One would expect little of the relevant antigenic protein to exist in the healthy plant; on the supposition that gene expression in gene-for-gene systems is autoregulated, one would expect the protein to be produced in quantity only after, and as a result of, infection. So, too, one would expect little of the relevant protein to exist in the uredospores; it would only be produced in quantity by the actively parasitizing mycelium. The argument was given in Section 3.18, and need not be repeated here.

This experimental difficulty is probably the reason why, relative to many far less important topics, common antigenic determinants have not received as much attention as they should.

4.5 Alternative Hypotheses About Common Antigens

Two very different hypotheses can be formed to explain the purpose of common antigenic determinants. There is the mimicry hypothesis, that emphasizes similar surfaces and host defense. There is now the protein copolymerization hypothesis, that emphasizes complementary surfaces and parasite nutrition.

The mimicry hypothesis assumes that the closer the antigenic determinants of the parasite resemble those of the host, the less likely the parasite is to be recognized as a foreign body by the host, and the less likely are defensive responses to be elicited from the host. This hypothesis has been stated by Rowley and Jenkin (1962), Dineen (1963), and Damian (1964) for parasites of animals, and De Vay et al. (1972), among others, for parasites of plants.

For parasites of mammals, which actively respond to the presence of foreign antigens by producing antibodies, the mimicry hypothesis is reasonable. However, no system of active response by plant hosts to foreign antigens, *qua* antigens, is known; the literature of this, extending back over many years, has been reviewed by Chester (1933) and Matta (1971). In the absence of evidence for active response to antigens by plants, the case for the mimicry hypothesis in parasitism of plants is decidedly weak.

The protein copolymerization hypothesis, of which we regard common antigens as a secondary concomitant, has already been discussed at length. It could apply equally to parasites of animals, although nowhere in the literature does there seem to be an inkling of the connection between copolymerization and shared antigens. In the feeding of animal parasites, protein copolymerization may yet take its place along with phagocytosis and other recognized systems. Very probably, for animal parasites both mimicry and protein copolymerization occur, each for its separate reason.

4.6 Common Antigenic Determinants in Other Than Gene-for-Gene Systems

In known gene-for-gene systems the taxa involved are narrow. In the pathogen the taxa are races within a species or form. In the host the taxa are usually varieties within a species or genus.

Another and related feature of known gene-for-gene systems is that susceptibility in the host can be changed to resistance, or vice versa, and avirulence in the pathogen can be changed to virulence, or vice versa. For example, we know of a gene-for-gene system in wheat stem rust because there are both *sr* and *Sr* genes in wheat, and both avirulence and virulence in the fungus. The balance between susceptibility and resistance, or between avirulence and virulence can be tipped one way or the other with relative ease compared with most host–pathogen systems. This, on the protein copolymerization hypothesis, traces back to a loose association between protein from the host and protein from the pathogen; i.e., it traces back to a high dissociation constant of the copolymer. Probably only point mutations are needed for change. A change of amino acid to one with a more hydrophobic side chain could reduce the dissociation constant enough to change resistance to susceptibility or avirulence to virulence.

Suppose now that the relevant protein from the host and the protein from the pathogen were more tightly bound, i.e., suppose that the dissociation constant of the copolymer was small. (Some dissociation constants of protein copolymers are very small. For example, the dissociation constant of the trypsin-pancreatic trypsin inhibitor copolymer, a dimer, is about 10^{-13}.) Host and pathogen would then be permanently compatible, in so far as protein copolymerization determines compatibility. Small changes, such as through point mutation, would not be enough, on this supposition, to cause dissociation. Drastic changes, such as through deletions, would almost certainly be forbidden.

Whereas gene-for-gene systems, often involving countless pathogenic races, reflect (on the protein copolymerization hypothesis) a relatively large dissociation constant, a more stable host/pathogen system, involving susceptibility of species, genera, or higher taxa, would require a relatively small dissociation constant.

Charudattan and De Vay (1972) studied antigenic relationships between four varieties of cotton *(Gossypium hirsutum)* and isolates of the cotton pathogens *Fusarium oxysporum* f.sp. *vasinfectum* and *Verticillium albo-atrum*. An antigenic determinant common to all four varieties of cotton and to all isolates of these fungi was found, despite the fact that some of the cotton varieties were resistant and some susceptible, and that some of the isolates of the pathogens were avirulent and some virulent. The tests, so far as they went, showed a relationship between *G. hirsutum* as a species and *F. oxysporum* f.sp. *vasinfectum* and *Verticillium albo-atrum* as species or formae speciales, and did not reflect differences within the host species or the pathogen form or species. The level of taxa involved in the interactions was higher than in known gene-for-gene systems. Similar results were obtained by Venkataraman et al. (1973). They too found shared antigens common to both resistant and susceptible lines of cotton, and to both avirulent and virulent isolates of *F. oxysporum* f.sp. *vasinfectum*.

The other side of specificity was clear too. Cotton did not have antigenic determinants in common with *F. moniliforme* and *V. nigrescens* which do not attack cotton. *F. oxysporum* f.sp. *vasinfectum* and *V. albo-atrum* did not have antigenic determinants in common with plants which, unlike cotton, they do not attack.

To return to the question why the antigenic relation between cotton and *F. oxysporum* f.sp. *vasinfectum* and *V. albo-atrum* was not disturbed by the resistance or susceptibility of the cotton varieties or by the pathogenicity or nonpathogenicity of the fungus isolates, Charudattan and De Vay cited the work of Bugbee and Sappenfield (1968) and Garber and Houston (1967). These workers found that the fungi colonize the cortex of the root, but resistance is not manifested by the host until the vascular system is invaded. The differences in resistance between varieties of cotton and in pathogenicity between isolates of the fungi are differences expressed in the vascular tissue. The antigenic relationships that ignore these differences are relationships determined by the cortical tissue.

Here we come back to the importance of a surviving nucleus in protein copolymerization (Sect. 2.4) and by implication in antigen sharing. The cortical cells have nuclei; the xylem has not. The cortex can be involved in a system of common protein antigens; the xylem cannot. Mass inoculation, purely an artifact, may disintegrate cortical cells. However, as Garber and Houston (1966) showed for *V. albo-atrum* in cotton, when the invasion of the cortical layers is limited to a few hyphae, the cell protoplasts survive, at least for a while. Hyphal strands develop both intercellularly and intracellularly. Sometimes they develop tangentially and intercellularly for several cell layers, and then penetrate directly into the interior of the cell; and when cells are penetrated, appressoria are common.

The part played by the xylem parenchyma is not yet clear. Another disease to show common antigenic determinants at species level is that caused by *Ophiobolus graminis* in wheat and oats (Abbott, 1973). Irrespective of whether the fungus isolates were pathogenic or not, he found a single precipitin band in agar gel diffusion tests when antisera of wheat and oat roots reacted with antigen preparations from *O. graminis*. The antigens were saline-soluble proteins. One assumes that the dissociation constant of the protein copolymer was small enough to maintain copolymerization irrespective of changes at subspecies—isolate—level. Here too there is survival, at least temporarily, of the host cell nucleus. Much of the infection is at the root surface, and invaded cells remain alive long enough to form "lignitubers" at the point of entry.

A matter of nomenclature arises. There is a possibility—we need not here try to evaluate the probability—that in the examples discussed in this section the relevant antigens of host and pathogen are monogenically determined. Compatibility would then be on a gene-for-gene system. The system would, however, differ from Flor's because it concerns susceptibility. In Flor's system it is the gene for resistance, not susceptibility or compatibility, that is at issue. ("For each resistance gene in the host there is a matching and reciprocal gene for pathogenicity in the fungus.") To avoid confusion it would seem advisable, for the present at any rate, to restrict the term gene-for-gene and use it only sensu Flor.

4.7 Common Antigens and Male–Female Recognition in Plants

Common antigens feature in pollination, and for much the same reason as in matters of host–pathogen recognition. A leaf is exposed to spores of countless species of fungi and to other inocula. Only a minute fraction of the total is able to infect; that is what host–pathogen recognition is about. A stigma is exposed to pollen grains of countless species. Only a minute fraction of the total is able to fertilize; that is what pollen–stigma recognition is about. Pollination and infection have common features; and pollination theory (Clarke et al., 1977) has recently almost caught up with infection theory in the matter of common antigenic determinants.

When pollen is recognized at the stigma surface, a mixture of components is released from cavities in the pollen walls. These components flow over and bind to the stigma surface which is covered by a secreted sticky coat. The binding of specific recognition factors determines the success of a mating. Clarke et al. (1977) used immunochemical methods with rabbit antisera to components from pollen and stigma surfaces of *Gladiolus gandavensis*. They found that the major antigen from the stigma surface shows immunological identity, and has the same molecular weight as an antigenic component from the pollen wall. Absorption experiments showed that the cross-reacting antibodies could be removed from the antisera, leaving antibodies that react only with the immunizing antigen. It is possible, Clarke et al. conclude, that the antigens may be isozymes of hydrolytic enzymes common to both pollen and stigma.

We will not leave the matter there, and logic takes us from common antigens, which in themselves do not bind without an intermediary, to complementary surfaces, which can copolymerize and therefore bind directly. That is, we suggest that copolymerization occurs between monomers each with at least two associating surfaces; that this copolymerization allows surface–surface recognition and binding; that it also allows the pollen tube to feed during its passage through stigma and style; and that the pollen tube is essentially a plant parasite which, through convergent evolution, behaves much as does the mycelium of, say, *Puccinia recondita* in leaves of a congenial wheat host.

A likely difference might well be in the dissociation constant of the copolymer. In a gene-for-gene system the dissociation constant must necessarily be relatively large; in a pollen recognition system it is likely to be relatively small, with the monomers tightly bound together.

Chapter 5 Other Large Molecules in Relation to Gene-for-Gene Disease

5.1 Introduction

We are in this chapter still concerned with how variation in gene-for-gene diseases is stored chemically and recognized in host-pathogen associations. The general problem was stated in Chapter 2, using wheat stem rust for illustration. There are potentially more than a million phenotypes of wheat that differ in their reactions to *Puccinia graminis tritici*, and potentially more than a million phenotypes of *P. graminis tritici* that differ in their reactions on wheat. Axiomatically, all the relevant information is stored in the genes, i.e., in the DNA. However, the expression of the information and its manifestation as potentially more than a million races of stem rust is almost certainly not at DNA level. Molecular recognition and chemical interaction between constituents of the host and those of the pathogen occur between the products of the genes. But which products? In Chapters 2, 3 and, obliquely, 4 proteins were named as the products, susceptibility being seen as host protein–pathogen protein copolymerization. In this chapter we consider candidates other than proteins. This is done, not to challenge the protein copolymerization hypothesis, but to put the chemical storage and recognition of variation in proper perspective. This has not been done adequately before; and the very large literature of phytoalexins testifies to a widespread misconception about the storage of variation in molecules and its part in host–pathogen interaction.

The title of this chapter mentions large molecules in relation to gene-for-gene disease. Largeness is implicit, if, as is frequent in gene-for-gene disease, we accept that variation is large. Large molecules are needed to store information about large amounts of variation. The other necessary attributes of the molecules are the repetition and shuffling of molecular residues. Thus we find molecular storage of information in poly compounds: polypeptides, polynucleotides, polysaccharides.

Proteins—polypeptides—have already been discussed in some detail in relation to copolymerization. In this chapter we turn to nucleic acids, complex carbohydrates including glycoproteins, and protein in lectins. We do this to evaluate the likelihood of their being involved in the direct host molecule–pathogen molecule contacts that determine gene-for-gene systems. We conclude that the likelihood is small.

5.2 DNA

DNA stores all biological variation (the RNA viruses excepted), and it does this by virtue of being a polymer. It is a polymer of mononucleotides, each mononucleotide consisting of a purine or pyrimidine base, desoxyribose and phosphate. There are only four different kinds of base in DNA: adenine and guanine representing purines, and thymine and cytosine representing pyrimidines. Yet with only these four bases as variables, DNA can store information endlessly, because it is a polymer in which the sequence of bases can be stretched and changed endlessly.

In almost all plant diseases involving eukaryotes, the relevant DNA in both host and pathogen is contained within the nucleus, and is tightly linked to other DNA in the chromatin on the currently accepted model of a string of beads. There is no known mechanism whereby nuclear DNA of a eukaryotic pathogen could break loose and come into direct chemical contact with the nuclear DNA of the host. Because there can be no molecular interaction at a distance beyond the range of chemical bonds, we exclude, for eukaryotes, the likelihood of a direct host DNA–pathogen DNA recognition and association.

5.3 RNA

For host–pathogen association at molecular level, RNA is a much more promising candidate than DNA. It shares DNA's ability to store biological variation, and, like DNA, does this by virtue of being a polymer in which the sequence of four bases can be changed endlessly. The replacement of thymine by uracil as the pyrimidine concerned does not affect the particular issue of the chemical storage of variation. Where RNA differs from DNA relevantly to our immediate topic is that it is mainly cytoplasmic rather than nuclear, and that the strands are shorter and freer to combine across a host–pathogen boundary.

The essential difficulty in finding a role for RNA in molecular contact between host and pathogen is the chemical invariance of the nucleic acids. Mutations occur, the mutations are stored within the RNA polymer as a change in the order of the bases, but the chemical reactivity of the RNA is essentially invariable. To pinpoint the problem, the essential difficulty is to explain how a change from the two hydrogen bonds associated with adenine and uracil to the three hydrogen bonds associated with cytosine and guanine could make susceptibility endothermic or resistance exothermic. Until this is explained, one must remain skeptical about the likelihood of finding a thermodynamic solution of the gene-for-gene relationship in host RNA–pathogen RNA associations. The chemical reactivity so conspicuous in the proteins is lacking in RNA.

The challenge is there: show how a host RNA–pathogen RNA association could make susceptibility endothermic and provide a basis for the feeding of the parasite that is the essence of parasitism. The RNA would have to be shown in a role other than just that of a messenger.

If one allows for RNA to be present as a messenger then gene-specific RNA is almost certain to be present both before and after infection. Demonstration that it is present demonstrates nothing that is not already to be expected. Howes et al. (1974) and Rohringer et al. (1974) used wheat stem rust complexes from which RNA was extracted. They obtained evidence for an active RNA component, possibly originating in the pathogen, and possibly a product of the gene for avirulence. Nothing in this suggests that the RNA was involved molecularly other than as a messenger. In gene-for-gene systems there is a specific DNA in the host which we call the gene for resistance or susceptibility, and a specific DNA in the pathogen which we call the gene for avirulence or virulence. In host–pathogen interactions, irrespective of whether they are compatible or incompatible, the genes are evidently turned on. Derived and specific RNAs are expected to be present, and the only cause for surprise would be a demonstration that they are not.

Another possibility is a hybrid RNA–protein association. Nucleic acid–enzyme associations play a large role in molecular biology; and there seems to be no reason why association could not occur across the host–pathogen boundary. Molecular association at RNA-protein level does not seem to have been suggested in the literature as a source of gene-for-gene interaction. It would have the advantage over pure nucleic acid associations in going half way towards the chemical reactivity of protein–protein associations. The challenge is there, to show how RNA-protein association in a gene-for-gene system could make susceptibility endothermic; but until the challenge is met, one must prefer the evidence for full protein–protein copolymerization.

5.4 Glycoproteins

5.4.1 The Nature of Glycoproteins

Glycoproteins are carbohydrates covalently bound to protein. Variation is stored both in the carbohydrate moiety and in the protein. During the past decade interest in glycoproteins has increased enormously, especially as components of animal membranes. In plant pathology interest has centered mainly on the possible role of glycoproteins as elicitors of phytoalexins and the hypersensitivity reaction, and this is the role that now concerns us.

In principle, the method of storing variation in complex carbohydrates is the same as in proteins and nucleic acids. Variation resides in the arrangement and identity of the oligosaccharide units, just as in nucleic acids it resides in the arrangement and identity of the nucleotide bases, and in proteins in the arrangement and identity of amino acid residues; but similarities must not be allowed to obscure differences. In passing from variation stored in proteins, the topic of Chapters 2, 3, and 4, to variation stored in carbohydrates, we pass from abundant variation derived from the proteins' isozymes to the much less abundant variation derived from the proteins' catalytic activity. There is a massive loss of variation through catalysis, because by definition isozymes are genetically specified variants that produce the same product by catalysis. The number of isozymes therefore measures the loss of variation through catalysis. It is an open question whether enough variation could survive catalysis to cope with the chemical storage of variation associated with, say, wheat stem rust. We must judge examples on their merits.

Glycoproteins exist both in solution and as constituents of membranes. It is with membrane glycoproteins that the relevant literature has been concerned. Membrane glycoproteins are found around cells, around the nucleus, and around the mitochondrion and other organelles.

5.4.2 Membranes and Membrane Glycoproteins

Ideas about the structure of membranes go back to the early work of Danielli and Davson (1935). Their model postulated a lipid bilayer, based on the calculation that the amount of lipid in an erythrocyte is about twice that needed to cover the entire surface. The hydrocarbon chains are orientated inwards and the polar heads outward towards the surrounding aqueous phase. The nonlipid components of the membrane are thought to be held by hydrophobic interactions similar to those discussed in Chapter 2, except that the reactions are now protein-lipid interactions in place of protein–protein interactions. In the fluid mosaic model of Singer and Nicholson (1972), protein and glycoprotein molecules are embedded in a mobile sea of lipid, two molecules thick. To consider just the glycoproteins, the attachment to the lipids is through hydrophobic residues of amino acids, to leave exposed the more hydrophilic part of the protein and the hydrophilic carbohydrate moiety. This leaves the membrane surface with a covering of carbohydrate chains extending into the surrounding aqueous phase. Winzler (1970) has reviewed the general literature. It is these carbohydrate chains that give the membranes much of their reactivity and variability.

There are also glycolipids in the membranes, with the lipid moiety embedded in the lipids of the membrane, and the carbohydrate moiety floating in the aqueous environment of the membrane. Presumably what holds for the behavior of the carbohydrate chains of glycoproteins holds with some similarities and differences for the behavior of the carbohydrate chains of glycolipids. However, the only relevant literature available is for glycoproteins; we must perforce ignore glycolipids, but should remember that they form a substantial part of the complex carbohydrates associated with membranes.

5.4.3 The Carbohydrate Determinants of the AB0 Blood Cell Types

The oldest and best known example of the storage of variation by carbohydrate residues is in the AB0 system of human blood group antigens. The A, B, and 0 types differ from each other by the terminal glycosyl residue on the oligosaccharide moeity of a membrane glycoprotein. If the terminal glycosyl residue of the oligosaccharide chain is N-acetylgalactosamine, the cell is of type A. If it is galactose, the cell is of type B. If both these residues are lacking, that is, if the oligosaccharide chain is shorter by one residue, the cell is of type 0.

5.4.4 One Gene–One Glycosidic Linkage Hypothesis

The relation, if there is a relation, of the complex carbohydrates to the gene-for-gene hypothesis might be seen in the one gene–one glycosidic linkage hypothesis of Roseman (1970). The heterosaccharide chains are assembled, residue by residue, by glycosyl transferases. Each glycosyl transferase is specific for the transfer of a particular sugar; and each glycosyl transferase is coded by a single gene. The heterosaccharide chains are assembled step by step, the transfer monogenically of a particular donor sugar creating a new acceptor molecule which is recognized by a second gene, and so on until the chain is complete.

5.4.5 Glycoproteins and the Elicitor Hypothesis

Molecules produced by pathogens which cause a hypersensitive response in their hosts have been called elicitors by Keen et al. (1972). These authors found that by inoculating a resistant variety of soybean *(Glycine max)* with *Phytophthora megasperma* var. *sojae* there was an increased production of phytoalexins. There has been considerable experimental work in recent years on this disease, and a hypothesis has been published to relate elicitors and the gene-for-gene system. The elicitor studied is a glycoprotein, which makes the literature relevant to this chapter.

Albersheim and Anderson-Prouty (1975) have stated the hypothesis in detail. They propose that on the surface of pathogens there occur glycoproteins which are the elicitors. The elicitors (they assume) vary in structure from race to race of the pathogen. For a race to be virulent and able to attack the host, the elicitor must have a structure not recognized by the host. Structural variations between elicitors of different races may result from the substitution of one glycosyl residue for another, from changes in anomeric configuration, in the addition or loss of glycosyl residues, or from the presence of other chemical moeities, such as methyl esters, and acetate and phosphate esters. If the variations are to be effective, they must control the process whereby pathogen and host recognize each other, but they must not greatly reduce the fitness of the pathogen. They therefore propose that the elicitors of the races of a gene-for-gene pathogen possess a basic structure common to all races, but which is made unique for each race by some change in the detailed structure of the glycoprotein.

Essential to the hypothesis of Albersheim and Anderson-Prouty is surface interaction between cells of pathogen and host. They believe that for varietal specificity to occur in a gene-for-gene system there must be an interaction between a glycoprotein on the pathogen's surface and a plasma membrane protein in the host. This interaction, they suggest, would elicit a defense mechanism in the host that would be specific for the particular host–pathogen combination.

5.4.6 An Active Glycoprotein from *Phytophthora megasperma* var. *sojae*

Following up the work of Keen et al. (1972) on *Phytophthora* stem and root rot of soybeans, Ayers et al. (1976a, b, c) and Ebel et al. (1976) studied the polysaccharide portions of highly purified preparations of the elicitor from both extracellular media and mycelial walls of *P. megasperma* var. *sojae*. The elicitor was determined to be predominantly a 3-linked glucan. It is biologically very active, activity being detectable when as little as 10^{-13} moles are applied to soybean cotyledons. It is this glucan which, the experimenters believe, is bonded to protein in the host cell walls, to make the host respond by producing a phytoalexin that causes resistant soybean varieties to be resistant.

There are two major objections to this hypothesis.

5.4.7 A Thermodynamic Objection

The essence of the hypothesis is that glycoprotein from the fungus is bound to protein in the host cell wall to produce a protein–glycoprotein complex which through stimulating phytoalexin production is the source of resistance. Resistance in gene-for-gene disease is exothermic (susceptibility being endothermic). The hypothesis therefore requires the production of the protein–glycoprotein complex to be exothermic. On the scant and incomplete evidence available, the production of protein–glycoprotein complexes appears to be endothermic; the hypothesis is contradicted by the available thermodynamic evidence.

The lectins concanavalin A and ricin are proteins that bind with glycoprotein receptors, concanavalin A specifically with glucose and mannose residues, and ricin with galactose residues (see Sect. 5.6). Kaneko et al. (1973) studied the effect of temperature on the agglutination of rat ascites hepatoma cells by these two lectins. Agglutination occurred strongly at 25° C but very weakly or not at all at 0° C. Raising the temperature from 0 to 25° C caused cells treated with concanavalin A and ricin to agglutinate rapidly.

Moscona (1961) set out to explain how cells join with one another to form tissues and organs of multicellular organisms. He measured the effect of rotation on the aggregation of dissociated cells. Culture flasks with equal aliquots of cell suspensions were rotated at 70 rpm at 38, 30, 25, and 15° C. After 24 h there was good aggregation at 38, and less at 30° C. At 25° C most cells remained free in suspension, and those that aggregated formed only minute clusters. At 15° C no aggregation occurred. These seem to be equilibrium conditions, because there was no material change of aggregation when rotation was extended from 24 to 48 h.

These results do no more than hint, in the absence of precise thermodynamic studies, that attachment to surface membranes and therefore probably the carbohydrate moieties of glycoproteins is endothermic. They hint that if the hypothesis of Albersheim and Anderson-Prouty (1975) were correct, resistance should be an endothermic process. In fact, in gene-for-gene disease, it is an exothermic process, susceptibility being endothermic (Chap. 3).

5.4.8 Elicitors Are Unspecific

The evidence from the work of Ebel et al. (1976) is that the elicitors from *Phytophthora megasperma* var. *sojae* are neither race-specific nor cultivar-specific. The elicitors reflect neither the difference between virulent and avirulent races of the pathogen, nor the difference between susceptible and resistant varieties of the host. There is no evidence to connect the elicitors with genes for virulence or avirulence in the pathogen, or genes for resistance or susceptibility in the host. That is, there is no evidence that the carbohydrate moeities of glycoproteins could be the molecular determinants of gene-for-gene systems.

There is a great conceptual difference between involving proteins (and nucleic acids) on the one hand, and carbohydrates on the other in host–pathogen molecular associations. On the classic one gene–one enzyme concept, it can be inferred with a high degree of probability that the genetic difference between a resistant and a susceptible host variety will be carried through to the coded proteins of the host, and that the genetic difference between an avirulent and a virulent race of the pathogen will be carried through to the coded proteins of the pathogen. However, no such inference can be made for carbohydrates. Relatively few of the amino acids of an enzyme are in contact with the substrate for binding and catalysis, and relatively few of the enzymes of host and pathogen are glycosyl transferases. The link between genetic differences and carbohydrate differences is slender; and what can be accepted a priori with reasonable safety about a gene–protein relation cannot be accepted without direct experimental evidence about a gene–carbohydrate relation.

5.5 Polysaccharides

Cellulose derivatives of *Xanthomonas campestris* and *X. phaseoli* have been studied by Morris et al. (1977), and some work on polysaccharides has been reviewed by Preston (1977).

Cellulose is most familiar to us as an insoluble structural compound, part of the cellular plant skeleton. Soluble extracellular cellulose derivatives are, however, also known; and those from *Xanthomonas*, called xanthans, are thought by Morris et al. to be involved in host–pathogen recognition.

The primary structure of the xanthans consists of a cellulose backbone, substituted on alternate glucose residues with trisaccharide side chains. The repeating unit is thus a pentasaccharide. The trisaccharide side chains fold down to align themselves with the main chain, and presumably stabilize the conformation by noncovalent interaction.

A clue to the function of xanthans is given by the observation that when mixed with even low concentrations of galactomannan, the mixture may gel although neither polysaccharide does so on its own. The proportion of mannose in the galactomannan is important, and the ratio of mannose to galactose must exceed 2:1 for gels to form. It is therefore deduced that noncovalent binding occurs between the unlike chains of xanthan and mannan; this may allow recognition at molecular level between host and pathogen. The interaction is not confined to galactomannans, but occurs also with other polysaccharides, including cellulose derivatives, with the β-1, 4-linked backbone. The walls of the host plant may therefore contain many polysaccharides to which the xanthan can attach itself and bind host and pathogen, noncovalently, at molecular level.

The hypothesis of Morris et al., that there are in the walls of host plants a number of polysaccharides to which bacterial polysaccharides might become attached, might well be correct as a general phenomenon, without bearing on gene-for-gene interaction. The *Xanthomonas* sp. listed by Day (1974) as being involved in a gene-for-gene system is *X. malvacearum* in *Gossypium*, and Brinkerhoff (1970) recognizes at least 16 different genes for resistance in this host plant. One must therefore allow for 2^{16} or, approximately, 60000 potential variations. There is nothing so far revealed about the chemical constitution of xanthans which suggests that they are able to store variation on this scale.

Even if xanthans play a role in pathogenesis by *Xanthomonas*, that role seems unlikely to be that of molecular determinants of variation in a gene-for-gene system; and gene-for-gene systems are what this chapter is about.

5.6 Lectins

Lectins are proteins or, occasionally, glycoproteins that can bind with sugar groups. They have been implicated in the binding of *Rhizobium* cells specifically to their leguminous symbionts; and because the *Rhizobium*–Leguminosae association is thought to be on a gene-for-gene basis (Nutman, 1969), they are relevant to this chapter. The evidence that lectins bind on a gene-for-gene system is, however, unconvincing.

Lectins (the name comes from *legere*, to choose) were first noticed as plant proteins able to clump red blood cells. From this come the older names hemagglutinins or phytohemagglutinins if they originate in plants. These synonyms are still used in the literature. Lectins are common in plants, especially legumes, and may account for as much as 3% of the protein in a plant; they also occur in animals.

Each molecule of a lectin has two or more sites, possibly grooves or clefts, into which complementary molecules of sugars or other oligosaccharides fit. These combining sites are where the lectins attach themselves to the saccharide chains of the membrane glycoproteins and glycolipids that surround the cells. So far as is known, lectins are specific for saccharides, and have not been found to combine specifically with any other compounds.

Specificity of lectins is determined by terminal sugars or several sugar units of an oligosaccharide. The lectin concanavalin A binds with glucose or mannose residues. The lectin from *Glycine max* binds with N-acetylgalactosamine, galactose and lactose residues. The sugars able to bind to lectins fall into four classes, depending on the C3 and C4 configuration of each monosaccharide or of a nonreducing aldopyranose portion of an oligosaccharide (Toms and Western, 1971). Because of this, lectins have specificity in binding, but limited specificity. It is this combination of specificity and limitation of specificity that makes lectins so useful in research. The limitation of specificity, as compared for example with the finer specificity of antigen–antibody systems, is important. Thus concanavalin A, besides whatever natural function it has in *Canavalia ensiformis* in which it is found, can agglutinate such diverse items as red blood cells, malignant cells, or glycogen, the common feature being the presence of glucose or mannose residues. Without this diversity, which implies limited specificity, concanavalin A or other lectins would probably never have been recognized for what they are.

There are several difficulties in associating lectins with gene-for-gene systems.

1. The experimental evidence is contradictory. The association of specific lectins with *Rhizobium* in Leguminosae was examined by Bohlool and Schmidt (1974). They found that the lectin from *Glycine max* combined with most strains of *R. japonicum* nodulating *G. max*, but not with strains of any *Rhizobium* species unable to nodulate *G. max*. From this they suggested that *Rhizobium* specificity was determined by an interaction between legume lectins and distinctive polysaccharides on the cell surfaces of *Rhizobium* strains. This finding has not been generally accepted. Law and Strijdom (1977) tested lectin-*Rhizobium* relations on a larger scale than Bohlool and Schmidt. They could not confirm Bohlool and Schmidt's finding, nor did they find in the lectins any specificity in the binding by strains of *Rhizobium* spp. that could explain the specificity of nodulation.

2. The variation that can be stored chemically by the four sugar groups is inadequate to match the variation inherent in gene-for-gene systems. This chapter centers around the proposition that to store large amounts of variation, large molecules are needed. The molecules involved in lectin agglutination are indeed large; but only a small fraction of the total glycoprotein or glycolipid molecule is involved in binding, and variation of the few saccharide residues is all the variation available. It is not enough. So, too, the lectin molecule is large, but the sites binding the saccharides are small and few. Thus, *Glycine max* lectin, a glycoprotein, has a molecular weight of 120000, but only two binding sites.

3. Lectin binding is a genetic improbability in a gene-for-gene system. At least four molecular surfaces are needed for lectin binding, thus, with a slant line indicating a surface, (host saccharide)//lectin//(pathogen saccharide). It is immaterial here whether the lectin comes from the host or pathogen. Changes in a gene-for-gene system, if lectins were involved, could be effected by changes in one or more of four genes. The gene-for-gene concept is, however, as the name implies, basically a two-gene concept when a new resistance gene in the host is matched by a corresponding mutation to virulence in the pathogen. There is no evidence to implicate a third or fourth gene in this simple matching. Indeed, such evidence, if found, would undermine the whole gene-for-gene hypothesis; and the inclusion of lectins in a molecular explanation of this hypothesis seems to be both an unnecessary and improbable complication.

Lectins may well play a part in other than gene-for-gene systems, e.g., in potatoes against *Pseudomonas solanacearum* (Sequeira and Graham, 1977). This matter is not at issue.

5.7 Phytoalexins

For the past two decades laboratories around the world have been trying to explain host–pathogen interactions in terms of phytoalexins. According to the phytoalexin hypothesis, the infection of a resistant plant is followed by an accumulation of substances called phytoalexins toxic to the pathogen in concentrations great enough to stop its growth. What concerns us here is that the identified phytoalexins have relatively low molecular weights and fixed molecular constitutions that cannot explain the presence of pathogenic races by the thousand. (One thinks here for example of the eleven known *R* genes and potential 2048 races of potato blight for which the phytoalexin hypothesis was first proposed.) To the best of my knowledge, no proponent of the phytoalexin hypothesis has even faced the challenge of explaining how phytoalexins could store variation chemically on the required scale. The onus to do so is clearly on them.

Those who believe that elicitors (Sect. 5.4) provoke the plant to produce phytoalexins simply throw the problem back to looking at the heavy molecules of DNA, RNA, proteins, and complex carbohydrates as possible candidates for elicitors and storers of chemical variation.

Phytoalexins should find an important place in a book on the healing and protection of wounds against infection, but are out of the main stream of our discussions. The wounds most pertinent to the large literature are the dead cells that result from the hypersensitive response of the host to invasion by an incompatible parasite. This was clearly foreshadowed in the well-known experiment of Müller and Börger (1941), in which previous inoculation of the cut surface of a potato tuber with an incompatible race of *Phytophthora infestans* inhibited secondary infection by a compatible race of this fungus or by *Fusarium caeruleum*. Vanderplank (1975) has discussed phytoalexins as preformed, localized antifungal compounds.

Chapter 6 Population Genetics of the Pathogen

6.1 Introduction

Plant breeders introduce resistance genes into crop plants. Whether the introduction will be successful or not depends on the pathogen. If it can match the resistance genes, i.e., if matching virulence alleles become prevalent in the population of the pathogen, the work of the plant breeder is largely undone. This is an old story in plant pathology. It is retold here simply to make the point that the decisive factor in breeding for resistance is the population genetics of the pathogen. The genetics of host resistance, which has received far more attention in the literature, is one step removed from the core of the problem. Resistance genes in the host influence the durability of resistance only indirectly, through their influence on the population genetics of the pathogen. In vertical resistance an influence is present. In horizontal resistance it is absent. That is what Chapter 1 was about.

Chapter 6 continues the discussion of gene-for-gene diseases; and gene-for-gene systems are probably the main, if not the only, systems involved in vertical resistance. In the context of practical plant breeding for resistance, gene-for-gene diseases are important, first, because these diseases include many of the more destructive epidemic diseases that attack major crops like wheat, and, secondly, because breeding for vertical resistance against these diseases has been a major preoccupation of plant-breeding institutes.

As the title states, this chapter is about population genetics. It is about the frequency of alleles for virulence and avirulence, and about the directional and stabilizing pressures that determine the frequency. Molecules enter the chapter, because the population genetics of pathogens becomes more intelligible if it is given a molecular hypothesis in the background.

6.2 Directional Selection Towards Virulence

Directional selection is another name for adaptation. If a resistance gene is introduced to produce a new cultivar, and if the pathogen adapts itself to the new situation by mutating to virulence or by increasing the frequency of virulence genes already present in the population, that is directional selection towards virulence. Directional selection has been the downfall of many cultivars introduced for their resistance. It follows in gene-for-gene diseases from the relation that for every gene for resistance in the host there is a corresponding gene for virulence in the pathogen.

Directional selection pressure may be large or small. The pressure is likely to be small or even absent if inoculum enters the crop from an outside source, and there is little or no feedback of inoculum from the crop to the outside source. It is likely to be large if the pathogen perpetuates itself within the crop, and is independent of outside inoculum.

Phytophthora infestans is a pathogen that perpetuates itself within the crop. It lives in the summer in the foliage of potatoes, and in the winter in the tubers. The process of adaptation is undisturbed by any necessity for the pathogen having to go elsewhere for part of its life cycle, and in climates that favor *P. infestans*, the destruction of vertical resistance by directional selection is rapid. The history of potatoes with the gene *R* 1 in the United States is typical. Kennebec, Cherokee, and some less successful cultivars with this gene became commercially available in the late 1940s and early 1950s. At first they were regarded as practically immune from blight, which is equivalent to saying that populations of *P. infestans* in the relevant area (mainly the northeastern United States) had few, if any, genes for virulence on the gene *R* 1. Because of this high resistance to blight, the cultivated area of potatoes with the *R* 1 gene increased fast, and by 1954 accounted for 6.3% of the area of certified seed grown in Maine. The 1954 season was climatically very favorable to blight. There was more than 90% defoliation by blight of unsprayed *R* 1-types before September 4 at the experiment station on the Aroostook Farm, Presque Isle, Maine (Stevenson et al., 1955). Blight in these *R* 1-types had appeared early in the season, being present in widely separated fields as early as the beginning of July (Webb and Bonde, 1956). This early appearance suggested that *R* 1-types were already grown widely enough for matching inoculum to overwinter in substantial amounts. Webb and Bonde (1956) confirmed this. Of 15 cull piles of potatoes that they examined, 11 gave isolates of *P. infestans* virulent on *R* 1-types of potato; of the 56 isolates of *P. infestans* that they examined, 30 were virulent on *R* 1-types. Sporangia from infected cull piles are the chief source of inoculum for summer fields of potatoes in Maine (Bonde and Schultz, 1943, 1944); *P. infestans* in the *R* 1-types had become self-perpetuating. Directional selection had caused an adequate proportion of virulence genes to accumulate in the local population of *P. infestans;* resistance given by the gene *R* 1 in the potato was matched and therefore "lost"; and from then on it was necessary to protect *R* 1-types against blight by spraying with fungicides, as if no *R* gene were present.

What happened to the potatoes of Maine with the gene *R* 1 was typical of what is generally called the boom-and-bust sequence. 1950 and thereabouts was the boom period, when the *R* 1-gene was seldom matched by virulence in the local population of *P. infestans.* 1954 saw the beginning of the bust period, when the *R* 1-gene was abundantly matched. A boom-and-bust sequence is notoriously common in plant breeding for resistance against gene-for-gene diseases. It involves two things, first, directional selection towards accumulating virulence in the population of the pathogen, and, second, the perpetuation of the virulent population through the survival of the pathogen within the resistant cultivar, either directly or through a feedback system. Directional selection pressures should probably always be assumed to be present. Perpetuation need or need not be present; and limited scope for perpetuation is one of the ways in which the threat of directional selection can be countered.

6.3 Stabilizing Selection Against Virulence

Stabilizing selection, also called homeostasis, is the opposite of directional selection. It is resistance against change.

In our present context stabilizing selection implies a preference for the avirulence over the virulence allele at a locus in the pathogen. We may visualize the pathogen population as having initially a high frequency of the avirulence allele and a low frequency of the virulence allele at a particular locus. The plant breeder introduces into the host the matching gene for resistance; obviously he would not wittingly have introduced the new host gene if the virulence allele had been the common allele in the pathogen, because then the host gene would have been ineffective from the start, and this would have been obvious from field experiments. The introduction of the resistance gene into the host starts the process of adaptation in the pathogen to the new genetic environment; that process is directional selection. Before the resistance gene was introduced into the host, however, the allele for avirulence was, by hypothesis, the common, and therefore the fitter allele. So the process of adaptation, which is directional selection, involves replacing the fitter allele with the less fit allele. This counter adaptation is stabilizing selection.

An instance of stabilizing selection overcoming directional selection occurred with wheat stem rust in Canada. In the early 1950s there were severe epidemics of stem rust in the Canadian wheat fields. Then the resistant cultivar Selkirk was introduced in 1954, and for more than a decade it dominated the wheat fields until it was superseded in 1965, for reasons unconnected with stem rust, by the cultivar Manitou with a different set of *Sr* genes. During that decade Selkirk maintained its high resistance to stem rust in Canada, and has done so ever since. During that decade, directional selection would have been towards virulence on Selkirk, that is, towards adaptation by the parasite to the host. This adaptation, with the rare exceptions given in the next paragraph, never took place, and stabilizing selection maintained a population of *P. graminis tritici* in which the frequency of avirulence on Selkirk was practically 100%. Stabilizing selection towards maintaining avirulence on Selkirk counterbalanced directional selection towards adaptation through virulence.

The virulent races of *P. graminis tritici* able to attack Selkirk were 15B-3 (Can.) and 15B-5 (Can.). Green (1971 b) has summarized the findings of the wheat stem rust surveys in Canada from 1956 through 1969. Only in 1960 were the virulent races found; in the other 13 years avirulence was 100% frequent in the sampled population. Put in terms of numbers, a total of 3522 isolates were examined in the 14-year period; of these only six, or 0.2%, belonged to race 15B-3 or 15B-5.

Katsuya and Green (1967) and Green (1971 b) explain that races 15B-3 and 15B-5 were nonaggressive and could not maintain themselves in nature. This is equivalent to saying that stabilizing selection acted against virulence. It is a matter of choice of words. If, after introducing a resistance gene into the host population, all known isolates of the pathogen virulent on that gene can be described as nonaggressive or unable to maintain themselves or in any other way that implies relative unfitness within the pathogen population, the meaning is that stabilizing selection acts against virulence.

6.4 Nonallelic Virulence Interaction in Stabilizing Selection

Two of the genes in Selkirk for resistance to stem rust are *Sr*6 and *Sr*9*d*. These two genes are relevant to our narrative. Virulence on *Sr*9*d* is common in *P. graminis tritici* in Canada. Virulence on *Sr*6 is not uncommon; but in Canada combined virulence on *Sr*6 and *Sr*9*d* is rare almost to the point of complete absence. There is a nonallelic interaction that keeps the two virulences apart in Canada; and it is this nonallelic interaction that provides the stabilizing selection pressure that protects Selkirk from stem rust (Vanderplank, 1975).

Table 6.1 gives relevant data. It records for each year the percentage of wheat stem rust isolates in Canada virulent on the genes *Sr*6 and *Sr*9*d*, each separately. From these data it is calculated what percentage should have been expected to be virulent on *Sr*6 and *Sr*9*d*, combined, if selection had been neutral. Thus in 1970 9.3% of the isolates were virulent on *Sr*6 and 86.2% on *Sr*9*d*; whence $9.3 \times 0.862 = 8.0\%$ should have been virulent on *Sr*6 and *Sr*9*d*, combined, if the two virulences had occurred independently of each other. In fact, no isolates were found in 1970 virulent on both *Sr*6 and *Sr*9*d*. It was also so for each year except 1975 when three isolates out of 332 tested had the two virulences combined. Combined virulence on *Sr*6 and *Sr*9*d* was rare to the point of insignificance, with three isolates found out of the hundreds expected over the years. The data in the table are not altogether complete, in that some isolates were not tested against *Sr*6 or *Sr*9*d* or both. These missing data were relatively few, and could not have changed the findings greatly; and in any case one assumes that in the context of virulence on *Sr*6 and *Sr*9*d*, their omission was random.

It is clear from this evidence for 1970 through 1975 that the stabilizing selection against virulence on the wheat cultivar Selkirk was largely or wholly caused by a nonallelic interaction between the pathogen's genes for virulence on *Sr*6 and *Sr*9*d*. This confirms an analysis (Vanderplank, 1975) made for the earlier years, 1965, 1967, 1968, and 1969. It is this nonallelic interaction that we must study in more detail.

Table 6.1. Percent of total isolates of *Puccinia graminis tritici* in Canada virulent on stem rust resistance genes *Sr*6 and *Sr*9*d*, singly and in combination, in the years 1970 through 1975

Resistance gene(s)	Year[a]					
	1970	1971	1972	1973	1974	1975
*Sr*6[b]	9.3	31.2	16.8	11.3	11.1	14.2
*Sr*9*d*[b]	86.2	68.8	82.8	88.7	92.3	93.6
*Sr*6 + *Sr*9*d* expected[c]	8.0	21.4	13.9	10.0	10.2	13.3
*Sr*6 + *Sr*9*d* found[b]	0	0	0	0	0	0.9

[a] The total number of isolates was 204, 135, 282, 106, 429, and 332 for the six years, respectively.
[b] From the data of Green (1971a), Green (1972a), Green (1972b), Green (1974), Green (1975), and Green (1976a) for the years 1970 through 1975, respectively.
[c] Expected on the assumption that the combination of virulences on the resistance genes was selectively neutral, without directional selection on the one hand or stabilizing selection on the other.

6.5 The Environmental Effect in Nonallelic Interaction Between Virulence on Genes *Sr*6 and *Sr*9*d*

Combined virulence on genes *Sr*6 and *Sr* 9*d* has been rare or absent in Canada in all the years when tests have been made; but it is not rare in the south of North America.

Table 6.2 shows that combined virulence on genes *Sr* 6 and *Sr* 9*d* is very common on wheat in Mexico, and common somewhat erratically in Texas. North of Texas, in the hard red winter wheat region and the hard red spring and durum wheat region, it was rare in 1973 and 1974 but less rare in 1975 which (it will be remembered) is one of the very few years in which it was recorded in Canada, even in trace amounts.

P. graminis tritici overwinters in Mexico, Texas, Kansas, and Oklahoma (Rowell and Roelfs, 1971). From there it proceeds northwards after the winter to Canada. It is clear from a comparison of results from Mexico and Texas (Table 6.2) with those from Canada (Table 6.1) that there is a strong dissociation of virulences on genes *Sr* 6 and *Sr* 9*d* along the path northwards. Combined virulence on genes *Sr* 6 and *Sr* 9*d* is fully fit to survive in Mexico but not in Canada. The few isolates with combined virulence that have been able to reach Canada have been described as unaggressive and unable to maintain themselves (see Sect. 6.3).

What causes the dissociation of virulences on *Sr* 6 and *Sr* 9*d*? Why is the nonallelic interaction between these two virulences weak or absent in the south, but strong in Canada? Temperature is a likely factor. Canada in the summer when wheat ripens and stem rust prevails is hotter than Texas in the winter or

Table 6.2. Percent of total isolates of *Puccinia graminis tritici* virulent on stem rust resistance genes *Sr*6 and *Sr*9*d*, together, in North America in 1973, 1974 and 1975[a]

	Year		
	1973	1974	1975
Mexico	73	92[d]	83
Texas (south of 30° N)	65	20	(5)[e]
Texas (central)	49	0	[e]
Hard red winter wheat area[b]	3	4	11
Hard red spring wheat area[c]	4	3	18
Canada	0	0	1

[a] From data of Roelfs and McVey (1974, 1975, 1976), except for Canada. Data for Canada from Table 6.1.
[b] From north-central Texas to the Platte River in Nebraska.
[c] North-eastern Montana, North Dakota, eastern South Dakota, and western Minnesota.
[d] Fall crop.
[e] Too few isolates for satisfactory analysis.

spring, when stem rust overwinters. The annual movement northwards of *P. graminis* is a movement from overwintering sources in the south to full summer temperatures in the north. The spores of *P. graminis* that reach Canada start the generations of stem rust furthest from those that overwinter; and between overwintering in the south and summer infection in Canada each successive generation occurs in a rising trend of average temperature. If the stem rust population adapts itself to a change of environment with the advent of summer, temperature stands out as the most likely factor in the change; and in any molecular interpretation of adaptive changes, temperature must play a key role. We return to this after considering other evidence about nonallelic interactions.

6.6 Dissociation of Virulence Through Repulsion

In Canada, virulence on *Sr6* and virulence on *Sr9d* repel each other and dissociate within the *P. graminis* population. There are other examples of dissociation of this sort.

The most conspicuous dissociation in recorded wheat stem rust surveys is with the virulences on *Sr6* and *Sr9e*. Available data about virulence on *Sr9e* do not go so far back as those about virulence on *Sr9d*; but dissociation is again clearly evident. In Canada virulence on *Sr9e* has behaved like that on *Sr9d* in relation to virulence on *Sr6*, and no further evidence for dissociation need now be given for that country. For the United States, data are given in Table 6.3. Virulence on *Sr9e* alone is common; about two thirds of the isolates were virulent on this gene in the years 1973 through 1975. Virulence on *Sr6* is not uncommon; about one tenth of the isolates in the same period were virulent on this gene. However, despite the extensive sampling, combined virulence on these two resistance genes was not found. In round figures, 400 isolates with combined virulence

Table 6.3. Percent of total isolates of *Puccinia graminis tritici* in the United States virulent on stem rust resistance genes *Sr6* and *Sr9e*, singly and in combination, in the years 1973, 1974 and 1975

Resistance gene	Year[a]		
	1973	1974	1975
Sr6[b]	13	6	11
Sr9e[b]	65	68	77
Sr6 + *Sr9e* expected[c]	8	4	8
Sr6 + *Sr9e* found[b]	0	0	0

[a] The total number of isolates was 1304, 2609, and 2418 for the three years, respectively.
[b] From the data of Roelfs and McVey (1974, 1975, 1976).
[c] Expected on the assumption that the combination of virulences was selectively neutral, and the virulences were randomly distributed.

Table 6.4. Percent of isolates of *Puccinia graminis tritici* in Canada virulent on stem rust resistance genes *Sr9a*, *Sr9b*, *Sr9d* and *Sr9e*, singly and in the years 1970 through 1975

Resistance gene(s)	Year					
	1970	1971	1972	1973	1974	1975
Sr9a[a]	12.8	34.4	25.7	21.6	8.7	6.9
Sr9b[a]	12.8	34.4	26.0	21.6	14.8	6.4
Sr9d[a]	86.2	68.8	82.8	88.7	92.3	93.6
Sr9e[a b]					83.7	84.0
Sr9a+*Sr9d* expected[c]	11.0	23.7	21.3	19.2	8.0	6.5
Sr9a+*Sr9d* found[a]	0	0	0	0.9	(0.5)[d]	1.5
Sr9a+*Sr9e* expected[c]					7.3	5.8
Sr9a+*Sr9e* found[a, b]					0.5	1.5
Sr9b+*Sr9d* expected[c]	11.0	23.7	21.6	19.2	13.7	6.0
Sr9b+*Sr9d* found[a]	0	0	0	0.9	[d]	1.5
Sr9b+*Sr9e* expected[c]					12.3	5.4
Sr9b+*Sr9e* found[a, b]					0.5	1.5

[a] Data from the same sources as in the Table 6.1.
[b] Data for gene *Sr9e* not available for 1970 through 1973.
[c] Expected on the assumption that the combination of virulences was selectively neutral.
[d] Too few isolates for satisfactory analysis.

on genes *Sr*6 and *Sr*9*e* were to have been expected; none was found. Roelfs and McVey (1975) have already commented that a combination of *Sr*6 and *Sr*9*e* would have been resistant to all cultures identified in the United States.

Wheat combining the genes *Sr*6 and *Sr*9*e* would be resistant to stem rust in both Canada and Texas, whereas the genes *Sr*6 and *Sr*9*d*, combined, are effective in Canada, but not in Texas. On the supposition that temperature is the deciding factor in the dissociation of virulence on these genes, one infers that wheat with the genes *Sr*6 and *Sr*9*e* would stay resistant in less warm environments than are needed to maintain resistance in wheat with *Sr*6 and *Sr*9*d*.

The genes *Sr9a* and *Sr9b* can be substituted for *Sr*6 in tabulations with *Sr*9*d* or *Sr*9*e*. (Allelism or linkage of resistance in the host is irrelevant in the context of virulence in the pathogen.) The matter is one of such great practical importance in agriculture that some relevant data are again tabulated (Table 6.4). The table clearly shows how virulence on genes *Sr*9*a* or *Sr*9*b*, on the one hand, repels virulence on genes *Sr*9*d* or *Sr*9*e*, on the other.

6.7 Association of Virulence as an Indirect Consequence of Repulsion

Because (in Canada) virulence on genes *Sr*6, *Sr*9*a* or *Sr*9*b* mutually repels virulence on genes *Sr*9*d* or *Sr*9*e*, it follows that there should be an association of avirulences and virulences on genes *Sr*6, *Sr*9*a* and *Sr*9*b*, on the one hand, and on genes *Sr*9*d* and *Sr*9*e*, on the other.

Table 6.5. Number of isolates of *Puccinia graminis tritici* in Canada in 1972 against which *Sr6*, *Sr9a*, *Sr9b* and *Sr9d* were effective or ineffective[a]

Race	Effective genes	Ineffective genes	Number of isolates
C 17	6, *9a*, *9b*, *9d*		3
C 18	6, *9a*, *9b*	*9d*	31
C 22	*9a*, *9d*	6, *9b*	1
C 33	6, *9a*, *9b*	*9d*	161
C 35	*9d*	6, *9a*, *9b*	35
C 41	*9d*	6, *9a*, *9b*	3
C 42	6, *9a*, *9b*	*9d*	1
C 44	6, *9a*, *9b*	*9d*	5
C 46	6, *9a*, *9b*	*9d*	5
C 47	6, *9a*, *9b*	*9d*	1
C 48	6, *9a*, *9b*	*9d*	1
C 49	6, *9a*, *9b*	*9d*	1
C 51	*9d*	6, *9a*, *9b*	1
C 52	*9d*	6, *9a*, *9b*	5

[a] Data of Green (1972b).

Table 6.5 illustrates the associations and dissociations of virulences on *Sr*6, *Sr*9*a*, *Sr*9*b*, and *Sr*9*d* in Canada in 1972. It will be remembered from Tables 6.1 and 6.4 that in 1972 there were no isolates virulent on genes *Sr*6 and *Sr*9*d* in combination, or on genes *Sr9a* and *Sr9d* as if combined, or on genes *Sr9b* and *Sr9d* as if combined. Table 6.5 shows which genes were effective and which were ineffective against isolates of the various races. Thus, against race C18 genes *Sr*6, *Sr*9*a* and *Sr*9*b* were effective and gene *Sr*9*d* was ineffective; that is, race C18 was avirulent on genes *Sr*6, *Sr*9*a* and *Sr*9*b*, but virulent on gene *Sr*9*d*. The entries for many of the races are identical; the races are distinguished by avirulence or virulence on genes other than the four shown. (Thus, race C42 is avirulent on genes *Sr*8, *Sr*11 and*Sr*15, whereas race C44 is virulent on these genes.)

In Table 6.5, if a race is virulent on gene *Sr*9*d* it is avirulent on genes *Sr*6, *Sr*9*a*, and *Sr*9*b*. If it is virulent on genes *Sr*6, *Sr*9*a* or *Sr*9*b*, it is avirulent on gene *Sr*9*d*.

Virulence, not avirulence, decides the associations. Avirulence on gene *Sr*9*d* can associate with avirulence on genes *Sr*6, *Sr*9*a*, and *Sr*9*b*. There are two races in Table 6.5 of special interest as examples of this. Race C17 is avirulent on all four genes, *Sr*6, *Sr*9*a*, *Sr*9*b*, and *Sr*9*d*. Race C17 is similar to or identical with old race 56 which dominated the population of *P. graminis tritici* in Canada in the 1940s; it has considerable fitness on susceptible wheat varieties, but its occurrence is now curbed by the newer wheat cultivars that are resistant to it. In race C22, genes *Sr*9*a* and *Sr*9*d* fall together, in avirulence. There is no close genetic linkage between alleles for virulence on genes *Sr*9*a* and *Sr*9*b*. (Although in wheat genes *Sr*9*a*, *Sr*9*b*, *Sr*9*d*, and *Sr*9*e* are allelic or closely linked, in *P..graminis tritici* the corresponding genes for virulence are not allelic or closely linked.)

Gene *Sr*9*e* does not enter Table 6.5, because it did not enter the race formula numbers in 1972. More recent surveys show how the pattern for gene *Sr*9*e* usually follows the pattern for gene *Sr*9*d*.

From the association of virulences on genes *Sr*6, *Sr*9*a*, and *Sr*9*b*, it follows that combined virulence on any two or three of these genes is found more frequently than would be expected if the virulences were randomly combined. In effect, where such close association occurs, there is little reason for wheat breeders to combine these genes in wheat cultivars; where conditions such as those reflected in Table 6.5 prevail, a combination of, say, genes *Sr*6 and *Sr*9*b* would give no more protection to wheat than gene *Sr*6 alone.

6.8 Evidence for Nonallelic Virulence Interactions in Some Other Diseases

In stem rust of oats caused by *Puccinia graminis avenae* there is, in both the United States and Canada, evidence that virulence on the resistance gene *Pg*8 interacts with virulence on *Pg*9. In western Canada virulence on *Pg*8 is frequent and sometimes 100% in sampled populations; virulence on *Pg*9 is correspondingly rare or absent. This accords with evidence for a nonallelic interaction, because if two virulences repel each other strongly, high frequency of virulence at one locus is likely to mean low frequency of virulence at the other. However, to reduce the risk of spurious correlations with data from widely different populations, we shall confine the analysis to data from eastern Canada where each of the two virulences occurs with substantial frequency. The number of samples tested in the east was relatively small, so the data for 1971 through 1975 are pooled. The data are those of Martens and Anema (1972), Martens (1972, 1974, 1975), and Martens and McKenzie (1976) for the five years, respectively. For the five years combined, there were 105 isolates tested. Of these 19.0% were virulent on resistance gene *Pg*8 and 76.2% on resistance gene *Pg*9. The percentage of isolates expected to be virulent on both *Pg*8 and *Pg*9, if virulence were randomly distributed and selectively neutral, is 14.5%. Only 1% was found.

In leaf rust of wheat caused by *Puccinia recondita* there is evidence in Canadian data for a nonallelic interaction of virulence on resistance genes *Lr*17 and *Lr*18. Relevant data are given in Tables 6.6 and 6.7. Table 6.6, drawn up on the same lines as Tables 6.1, 6.3, and 6.4 for *P. graminis tritici*, shows that the percentage of isolates found with combined virulence on genes *Lr*17 and *Lr*18 is significantly less than the percentage expected on the assumption that the combination of virulences was selectively neutral. Table 6.7, drawn up on the basis of effective and ineffective resistance genes, on much the same lines as in Table 6.5 for *P. graminis tritici*, brings out clearly how virulence on gene *Lr*17 and virulence on gene *Lr*18 repel each other. Avirulence at both loci is common, but virulence at one locus almost always means avirulence at the other.

Wolfe et al. (1976), citing unpublished work by Priestley, give evidence for a nonallelic virulence interaction in *Puccinia striiformis* which causes stripe rust in wheat.

In *Puccinia*, then, occasional nonallelic virulence interaction is found, and this gives special importance to the matching resistance genes in the host. In *Phytophthora infestans* the situation is not clear. Potato breeders have shown special

Table 6.6. Percent of total isolates of *Puccinia recondita* in Canada virulent on leaf rust resistance genes *Lr*17 and *Lr*18, singly and in combination, in the years 1971 through 1975

Resistance gene(s)	Year[a]				
	1971	1972	1973	1974	1975
*Lr*17[b]	7.1	5.9	4.3	8.4	1.8
*Lr*18[b]	23.1	23.1	34.2	36.9	9.5
*Lr*17 + *Lr*18 expected[c]	1.6	1.4	1.5	3.1	0.2
*Lr*17 + *Lr*18 found[b]	0	0.8	0	0	0

[a] The total number of isolates was 225, 169, 164, 179, and 327 for the five years, respectively.
[b] From the data of Samborski (1972a,b), Samborski (1974), Samborski (1975) and Samborski (1976).
[c] Expected on the assumption that the combination of virulences on the resistance genes was selectively neutral.

Table 6.7. Number of isolates of *Puccinia recondita* in Canada in the five years 1971 through 1975 against which genes *Lr*17 and *Lr*18 were effective or ineffective[a]

Effective genes	Ineffective genes	Number of isolates
*Lr*17	*Lr*18	244
*Lr*18	*Lr*17	55
*Lr*17, *Lr*18		766
	*Lr*17, *Lr*18	1

[a] Data from the same sources as those of Table 6.6.

interest in the resistance gene *R*1, but directional selection in the pathogen towards virulence on this gene has been so strong that it nullifies any effort to determine whether it is involved in nonallelic interaction. For many other pathogens relevant data are scarce, and it seems likely that it will take many years before the frequency of the occurrence of nonallelic virulence interactions can be satisfactorily assessed.

6.9 A Hypothesis About the Nonallelic Virulence Interaction

When in previous sections it has been said that virulence at one particular locus repels virulence at another, the reference is to repulsion in the population. As virulence at the one locus becomes more frequent in the population, virulence at the other becomes less frequent. It is not implied that there is chemical repulsion within the phenotype. On the contrary, it is implied that there is interference within the phenotype, and this means chemical contact.

Earlier chapters were given over to evidence that in gene-for-gene systems susceptibility of the host to the pathogen involves the copolymerization of protein. Much of the present chapter is given over to demonstrating a nonallelic interaction between virulences in *P. graminis tritici* on wheat stem rust resistance genes *Sr*6, *Sr*9*a*, *Sr*9*b*, *Sr*9*d*, and *Sr*9*e*, and a corresponding interaction in some other diseases. It would be consistent to suggest, as a hypothesis, that it is the relevant virulence proteins of the pathogen that are involved in the interaction, that they cause it through polymerization, and that the polymerization occurs because the relevant protein surfaces retain vestiges of their common ancestral form.

The requirements of this hypothesis are (1) that the nonallelic interaction, like protein polymerization, is favored by higher temperatures. The environmental evidence, such as it is, is compatible with this requirement (2) virulence, not avirulence, should determine nonallelic interaction. The supporting evidence, already given in detail in this chapter, is strong (3) there should be both complementarity and similarity of protein surfaces (see Sect. 4.3). Evidence for the coexistence of complementarity and similarity, as in Tables 6.4 and 6.5, is strong. Virulence on genes *Sr*6, *Sr*9*a* or *Sr*9*b* is complementary to virulence on genes *Sr*9*d* or *Sr*9*e*; and among themselves virulences on *Sr*6, *Sr*9*a*, and *Sr*9*b* behave similarly, as do virulences on *Sr*9*d* and *Sr*9*e*.

The fourth requirement (4), that the relevant protein surfaces should retain vestiges of a common ancestral form, needs elaboration. The tightness of protein polymerization depends largely on the area of the surfaces buried in polymerization, and this depends largely on the number of amino acids residues buried. A larger buried area means a smaller dissociation constant and a greater likelihood that the polymer would exist even at low biological temperatures. A small buried area is likely to mean polymerization only at high temperatures. This fits the facts of our problem, so far as they are known. There is nothing to suggest a nonallelic virulence interaction at low temperatures; if the hypothesis is correct, the buried protein surfaces must be small in area. This is what one would expect if the relevant protein surfaces diverged from a common ancestor, not sufficiently to develop entirely new surfaces, but sufficiently to leave only vestiges of common ancestry. So far as *Sr*9*a*, *Sr*9*b*, *Sr*9*d*, and *Sr*9*e* are concerned, it does not matter whether we regard them as alleles or closely linked duplicates; they can readily be assumed to have a common ancestor, and virulence on them to have a common ancestral protein surface. Even genes like *Sr*6, though not allelic or closely linked, may well have had a common ancestry; the alternative is to assume that all *Sr* genes, all *Lr* genes, etc. arose from independent sources, and this is not easy to accept.

Did all the *Sr* genes in the host and the corresponding virulence genes in the pathogen each begin with, or develop, a perfect individual identity and specificity, as Flor's gene-for-gene hypothesis must suppose, or do vestiges of some common ancestry persist, to become manifest only in certain conditions of, say, temperature? Flor's hypothesis must suppose that as resistance alleles at a locus or duplicate genes acquired separate identities and specificities, so too did the corresponding virulence genes in the pathogen; a few of these may still be in a transitional stage of multispecificity conditioned by small vestigial areas on the coded proteins; and, if this is so, smallness would make multispecificity occur only at appropriate temperatures.

This matter of identity raises another question. How is identity maintained? What prevents virulence proteins polymerizing among themselves? Presumably there is a block, possibly through steric hindrance. In the absence of any information about quaternary structure, there is little value in further discussion, except to point out that although head-to-tail polymers (as in Sect. 4.3) can be used to provide the simplest illustration, the actual quaternary structure could be, and is likely to be, very different.

6.10 The Second Gene-for-Gene Hypothesis

Flor's gene-for-gene hypothesis deals with gene identity and nothing more. In his work on the flax–flax rust system, he concluded that for each gene determining resistance in flax, there is a specific and related gene determining virulence in the rust fungus. The genes determining virulence in the fungus identify the genes determining resistance in the host, and vice versa.

What we have been discussing in this chapter shows that genes have qualities as well as identities. They differ in quality. Thus, in *Puccinia graminis tritici* virulences on genes *Sr*6, *Sr*9*a*, and *Sr*9*b* are roughly similar in behavior and different at appropriate temperatures from virulences on *Sr*9*d* and *Sr*9*e*, these two virulences being themselves roughly similar. These quality differences in virulence in the pathogen reflect quality differences in the resistance genes of the host, which are directly relevant to practical plant breeding.

Vanderplank (1975) formulated a second gene-for-gene hypothesis, with special reference to the quality of resistance genes in plant breeding and practical agriculture. In host–parasite systems in which there is a gene-for-gene relationship, the quality of a resistance gene in the host determines the fitness of the matching virulence gene in the parasite to survive when the virulence is unnecessary; and reciprocally the fitness of the virulence gene to survive when it is unnecessary determines the quality of the matching resistance gene as judged by the protection it gives to the host.

The second gene-for-gene hypothesis was proposed on evidence from potato blight and wheat stem rust. In populations of *Phytophthora infestans*, virulence on the gene *R*4 is abundant and almost universal even in potato fields without this resistance gene. That is, directional selection in favor of virulence is not substantially involved in this abundance. Because of this preexisting abundance of virulence the gene *R*4 is, and always has been, practically useless to potato breeders. In contrast with this, the resistance given by the gene *R*1 has been useful (at least temporarily) to potato breeders. Virulence on this gene did not preexist in populations of *P. infestans*, except in trace amounts, before potato breeders started to use the gene; it became abundant only as a result of directional selection after *R*1-types of potatoes were grown commercially (see Sect. 6.2). This difference between genes *R*1 and *R*4 illustrates the second gene-for-gene hypothesis. The genes *R*1 and *R*4 differ in quality and value to the potato breeder because their matching virulences differed in fitness to survive when they were unnecessary to *P. infestans*, i.e., when potato fields lacked these resistance genes. The gene *R*4 was called "weak", because its incorporation into a potato cultivar would have had only a weak effect on resistance to blight as judged in a farmer's field. The gene *R*1 was called "strong", because, in the days before populations of *P. infestans* had become adapted to it by directional selection, the incorporation of this gene into a potato cultivar had a strong effect on resistance as judged by the amount of blight in a farmer's field.

For wheat stem rust in Canada the message of Table 6.1 is that, when combined, the resistance genes *Sr*6 and *Sr9d* are strong because combined virulence on them is rare or absent. Together in Canada they provided, and still can provide, strong resistance against *Puccinia graminis tritici.* On its own, the resistance gene *Sr9d* is weak, because *Sr9d* without *Sr*6 gives only weak resistance, virulence on it being common in the absence of virulence on *Sr*6.

The quality differences recognized in the second gene-for-gene hypothesis extend not only to strong resistance genes but also, in the opposite direction, to ultra weak resistance genes. Table 6.5 shows that in Canada the gene *Sr9b* would be ultra weak in combination with *Sr*6, because to a wheat breeder its value in this combination would be even less than what one would expect from a random distribution of virulence.

What emerges clearly is that the strength or weakness of a gene can be judged only in relation to the genetic background. With *Sr9d* in the genetic background, *Sr*6 in Canada is strong; with *Sr9b* in the background, *Sr*6 is ultra weak. These opposite effects are clear in several other permutations and combinations of *Sr*6, *Sr9a*, *Sr9b*, *Sr9d*, and *Sr9e.*

The influence of the genetic background brings about some paradoxical situations. Canada obtains its inoculum of *Puccinia graminis tritici* from the United States. If by friendly international treaty it were possible for the United States to guarantee that it would export northwards populations of *P. graminis tritici* all of which were 100% virulent on gene *Sr9d*, Canada would benefit from this virulence by being able to control stem rust effectively by a single gene, *Sr*6. Evidence for this paradox that puts an agricultural premium on virulence (on *Sr9d*) is documented in Tables 6.1 and 6.5.

6.11 The Commonness of Weak Resistance Genes

By the concentration on a limited group of resistance genes, *Sr*6, *Sr9a*, *Sr9b*, *Sr9d*, and *Sr9e*, the impression might have been given in this chapter that measurable nonallelic interactions between virulences are common, and that most resistance genes are strong. This impression would be entirely false. Most resistance genes are weak (Vanderplank, 1968, 1975).

The same field surveys that have been discussed in the tables of this chapter provide data, not analyzed here, which suggest that most *Sr* genes other than the five listed in the previous paragraph are weak; and Osoro and Green (1976), in laboratory tests in Canada, have confirmed that genes *Sr* 7*a*, *Sr* 8, *Sr* 10, *Sr* 11, and *Sr* 15 are weak in the circumstances of their tests. So, too, weak genes predominate in relation to other cereal rusts and powdery mildews, and potato blight.

6.12 The Use of Weak Resistance Genes

Stabilizing selection against virulence operates weakly, or not at all, when the matching resistance gene is weak. That is contained in the second gene-for-gene hypothesis. This does not mean, however, that weak genes have no use agriculturally. They are indeed useful, but only when directional selection in favor of virulence operates weakly, or not at all. Stabilizing selection, however weak it might be, can then balance directional selection of equal weakness.

Wheat stem rust in Canada has been used to illustrate most of the topics of this chapter, so let us continue to use it. Make the realistic assumption, for argument's sake, that *P. graminis tritici* does not overwinter substantially in Canada, i.e., accept that *Berberis* spp. have been effectively eradicated. What then would be the consequence of releasing for agriculture a new wheat cultivar resistant to most races of *P. graminis tritici* arriving from the south? Would there, as a result of this release, be directional selection towards an increase of virulence on this cultivar? During the summer, virulent races could multiply and increase the relative frequency of virulence. This is directional selection; but it would mostly end with the end of summer. Any further maintenance of directional selection would require the survival of the extra virulence during a two-way journey over many thousands of kilometers. Spores would have to be blown southward to overwintering sites in amounts sufficient to change the population of *P. graminis tritici* that overwinters, and to maintain this change all through the northward sequence of generations in winter wheat and spring wheat back to Canada. Unless stem rust in Canadian wheat fields could significantly alter the virulence pattern in the remainder of North America that is relevant, there would be no significant directional selection; and no significant alteration can be traced in race surveys. For climatic reasons *P. graminis tritici* is forced to migrate, and so cannot cling to a new wheat cultivar in Canada; and in proportion to its divorce from the cultivar stabilizing selection is weak.

Case A, in the top half of Figure 6.1, illustrates very weak directional selection being balanced by very weak stabilizing selection. If directional selection is weak enough, weak resistance genes can cope with it, even though the stabilizing selection pressure they give is weak.

There are three requirements for the use of weak resistance genes in agriculture.

1. Directional selection must be weak.
2. Several different genes should preferably be used.

Simple arithmetic explains why. If, e.g., there are three resistance genes in a cultivar each separately giving resistance to 90% of the population of the pathogen, the three together will give resistance to 99.9% of the population. We assume

Directional selection
A ⇄
Stabilizing selection

Directional selection
B ⇄
Stabilizing selection

Fig. 6.1 A and B. For vertical resistance to be stable, directional selection for virulence must be balanced by stabilizing selection against virulence. (A) Directional selection is weak, stabilizing selection needs to be no more than correspondingly weak, vertical resistance has a good chance of succeeding, and its use can be recommended. (B) Directional selection is strong, stabilizing selection needs to be equally strong for vertical resistance to succeed, and the chance that this can be so is correspondingly small

in this calculation that the three relevant virulences in the pathogen's population occur randomly and do not interact; random occurrence without interaction of virulence is as good a way of defining weak resistance genes as any.

3. The genes must be properly deployed so as to isolate the cultivar epidemiologically. The histories of the great wheat stem rust epidemics in North America, in 1904, 1916, 1935, and 1953, show that there was a movement of inoculum northwards. Weak resistance genes can best help to break the movement if they differ from those present in cultivars grown further south.

6.13 The Gap in the Use of Vertical Resistance in Plant Breeding

Nothing has bedeviled plant breeding for resistance to disease more than the assumption that what holds for wheat stem rust in Canada, or other diseases and environments which favor the plant breeder, ought to hold for wheat stem rust in East Africa, potato blight, coffee rust, and other less favored diseases in less favored environments. At international congresses successes in breeding for resistance against disease are paraded by plant breeders working in circumstances in which directional selection is almost zero. These successes are exhibited as standards of what ought to be done elsewhere, and listeners return home to realities quite different. Success does not follow.

The argument is quantitative, and illustrated in Figure 6.1. At the top, in part A, directional selection is shown as weak, and easily countered by equally weak stabilizing selection. Vertical resistance, given the necessary resistance genes, is an appropriate and successful method of control. At the bottom, in part B, directional selection is shown as strong; and for vertical resistance to be stable and successful, stabilizing selection would have to be equally strong. The long line in part B representing directional selection would have to be matched by an equally long line representing stabilizing selection. Unfortunately the two lines are seldom equal in practice; stabilizing selection is inadequate; and vertical resistance gets a bad name, ignoring its evident usefulness, because it has been used where it should never have been used. Its use has been misguided.

6.14 A Guide to the Use of Vertical Resistance

Use vertical resistance to disease in plant breeding, when the circumstances are such that the disease is largely under natural control even when vertical resistance is absent. This guide seems to fit the facts well enough. It implies that vertical resistance has the power to improve a situation that needs only slight improvement, but cannot master a situation needing great improvement. As a generalization it seems true to say that most examples of the successful and stable use of vertical resistance are in regions where the relevant crops were successfully grown before there were organized institutional attempts to control disease by plant breeding and, in particular, by vertical resistance.

Consider stem rust in red spring wheat in North America. This choice of example is obvious, because it is from stem rust in spring wheat in North America that much of the literature of vertical resistance has come. Go back to the previous century. Wheat was the pioneers' crop in North America. It was grown so successfully that America was soon pouring wheat into Britain in such quantity and of such quality that market prices slumped to a 150-year low in 1894–1895. Clearly, the American industry was not an ailing one. Carleton (1899) made a survey of the rust situation and noted that rust attacked mainly late varieties. He had received 494 reports of attacks on late varieties against 25 of attacks on early varieties. Evidently rust arrived late, near ripening time. He commented that in North Dakota wheat does not commonly seem to be damaged, and, again, that in the Dakotas it is as a rule, too dry, cool, and breezy for rusts to do much damage. To come to this century, Waldron (1935), writing in North Dakota, stated that as a matter of fact in 50 years of North Dakota farming there had been only three first-class epidemics of wheat stem rust. They were in 1904, 1916, and 1935. Lesser epidemics, he noted, occurred in 1923 and 1927. The decade that ended in 1929 was the last in which common spring wheats were not protected by vertical resistance deliberately introduced by wheat breeders. Stakman and Fletcher (1930) have left a contemporary account of it that is supported by the early supplements of the *Plant Disease Reporter*. To Stakman and Fletcher stem rust in wheat and oats was primarily the result of infected *Berberis* bushes in the neighborhood. The great 1916 stem rust epidemic sparked off the *Berberis* eradication campaign, because it was obvious that much of the inoculum came from infected *Berberis* bushes, and Federal legislation for the destruction of *Berberis* began in 1918. Stakman and Fletcher observed the campaign, and gave example after example of the spread of stem rust from infected *Berberis* bushes into fields of wheat and oats with a background of little stem rust. *Berberis* infection is now largely a matter of history; what does concern us is the background of little stem rust. There was a concensus about low background infection; and nowadays when we read reports from Canada and the United States that in some particular year stem rust was rare, thanks to the wheat breeders, we should remember that all this about rarity was said long ago, before wheat breeders moved in. The picture painted nowadays of a rust-racked wheat industry waiting to be rescued by breeders is good propaganda but bad history.

The other side of the story is equally clear. Vertical resistance has been successful, and, when stem rust occurs, it occurs more in stem rust-susceptible wheat varieties than in those with vertical resistance. The purpose of the previous paragraph was not to dispute that vertical resistance was a gain, but to put that gain in perspective and, more particularly, to illustrate the guide given at the beginning of the section. Vertical resistance can be used to improve, but seldom to rescue.

The converse of this guide seems to be largely true, especially in relation to trying to substitute the use of vertically resistant varieties for control by fungicides. If a crop could be grown only with the lavish use of fungicides, it would be hazardous to dispense with the spray pumps and try to rely on vertical resistance. It is difficult to recall any example of vertical resistance replacing the heavy use of fungicides, except very temporarily.

The guide is for vertical resistance as defined in Chapter 1. There are well-documented examples of resistance rescuing a crop, but on examination of independent evidence, the resistance seems to be largely horizontal, or, if vertical, associated with unusually great pressure of stabilizing selection.

6.15 How to Improve the Performance of Vertical Resistance

Part B of Figure 6.1 illustrates the problem. To improve vertical resistance one must either reduce directional selection pressure or increase stabilizing selection pressure.

Directional selection is closely bound to the amount of disease. The more the disease, the greater is the selection pressure towards increased virulence. The relation is evident: Fields that are more heavily diseased contribute more heavily to the pathogen's population, and therefore can influence the frequency of virulence more. (The guide given in the previous section is a reflection of this relation.)

Anything that reduces the background level of disease improves the likelihood of vertical resistance being useful. A combination of horizontal resistance and vertical resistance makes vertical resistance more effective. So, too, vertical resistance is at its best when the climate is suboptimal for disease, or when anything in the environment (including fungicides) reduces disease. It follows also that vertical resistance is more apt for epidemic (sporadic) disease than endemic disease.

Stabilizing selection pressure is increased with long crop rotation (especially for soil-born diseases) or other measures that force the pathogen to leave the resistant cultivar or any other host plant bearing the relevant vertical resistance genes.

Stabilizing selection reflects gene combinations in the host. Special attention should be given to combinations of genes such as *Sr*6 and *Sr*9*d* or *Lr*17 and *Lr*18 in wheat or *Pg*8 and *Pg*9 in oats with matching virulences that (in appropriate climates and environments) dissociate more than randomly in the pathogenic population. This source of stabilizing selection has been generally missed by plant breeders. Most combinations leave the resistance genes weak, and effective only when the directional selection pressure is almost zero (see Sect. 6.12). However, in less favored agricultural circumstances, when stabilizing selection is a necessity if vertical resistance is to be durable, research into nonallelic virulence interactions needs strengthening. The emphasis in research at present is on resistance genes in the host: on the reaction types they give with selected pathogenic isolates, on their availability and freedom from undesirable linkage, and similar matters of direct interest to specialists in the host plant. At least equal emphasis in research is needed on genes for virulence in the pathogen and the behavior of combinations of virulence in the phenotype.

Resistance genes commonly have an additive effect on reaction type; several genes in combination can sometimes condition a highly resistant reaction type (e.g., a fleck reaction) that individually they cannot. No evidence is yet available to demonstrate that combinations of resistance genes which give the best reaction types are the combinations which give the greatest nonallelic interactions between the matching virulence genes and the greatest stabilizing selection pressure. There is no apparent reason why such a correlation should exist, and breeders' decisions made largely on the basis of reaction type may sometimes be unfortunate.

6.16 The Durability of Vertical Resistance

Boom-and-bust cycles have been documented so often for vertical resistance that there is a danger of the impermanence of vertical resistance becoming a dogma of plant pathology. In fact, there is no reason why vertical resistance, appropriately used, should not be stable. Figure 6.1 illustrates this. If the arrow representing stabilizing selection is as long as, or longer than, the arrow representing directional selection, vertical resistance will be stable. Admittedly, great stabilizing selection pressure is probably rare, and vertical resistance is likely to be stable only when used as guided in Section 6.14. With this restriction, it would seem better to aim at timelessness than at a program of constantly introducing new resistance genes into cultivars. An example repeated here for its relevance is the wheat cultivar Selkirk with stem rust resistance genes *Sr*6 and *Sr*9*d* (among others). From the time it was introduced a quarter of a century ago it has never been seriously threatened by stem rust in Canada (except, perhaps, briefly in 1960) because a nonallelic interaction has caused stabilizing selection to exceed directional selection. One wonders to what extent purposeful research into stabilizing selection could multiply such examples.

Even when absolute timelessness cannot be achieved, because directional selection pressure exceeds stabilizing selection pressure, the useful life of a new resistant cultivar can be increased by increasing stabilizing selection. The common concept of the breakdown of resistance in a new cultivar is that the pathogen has mutated to produce matching virulence. Accept this for argument's sake (although it is questionable whether when large acreages are involved the necessary virulence does not preexist in the pathogenic population at low frequency). Because most inoculum begins to die out after the season ends (if it did not, there would be little hope for vertical resistance), few mutations survive. Increased stabilizing selection means that fewer mutations survive; and a lower survival rate means greater durability of vertical resistance and a smaller turnover of "new" resistance genes needed in the host to block virulence changes in the pathogen. It would be good agricultural policy to spend on research on stabilizing selection funds commensurate with those now being spent on searching for "new" resistance genes; and in the long run the funds would serve the same purpose.

6.17 The Effect of Vertical Resistance on Epidemics

The course of an epidemic in a host population with resistance genes is determined (among other factors) by the frequency in the pathogen's population of virulence genes that can match the resistance genes. With host–pathogen specificity, when virulence and avirulence occur without intermediates, the reasoning is simple.

For illustration, suppose that a field of a cultivar is resistant to 99% of the spores reaching it, and susceptible to 1%. Resistance then has the effect of reducing the initial amount of disease to 1% of what it would have been in a susceptible cultivar, other things being equal, and this would delay the resulting epidemic by the period needed for disease to increase 100-fold. For the phase of exponential increase of an epidemic (and this is the phase most relevant to normal levels of initial disease) quantitative relations are easily established (Vanderplank, 1963, 1968). They serve a useful purpose in showing what issues are involved.

The equation of the relation is

$$\Delta t = -2.3(\log_{10}\alpha)/r$$

where Δt is the delay which vertical resistance brings about in the epidemic, α is the frequency (as a proportion) of relevant virulence in the pathogen's population, and r is the exponential (logarithmic) infection rate. Thus if 1/50 of the spores reaching a field are virulent (so that the frequency of virulence is 0.02 or 2%) and $r=0.4$ per day, the delay brought about by vertical resistance is $-2.3\ (-1.7)/0.4=10$ days, approximately.

The benefit from vertical resistance, measured as a delay in the epidemic, is inversely proportional to the infection rate. The benefit is therefore least when the weather favors disease, i.e., it is least when it is needed most. On the other hand, horizontal resistance or any other factor that reduces the infection rate increases the benefit. In plant breeding, vertical resistance is not something to be manipulated in isolation; it is always bettered by being used against a background of horizontal resistance.

When there is no host–pathogen specificity, in the sense that all isolates of the pathogen can attack all varieties of the host, vertical resistance can have the effect of reducing the infection rate, a matter discussed in Section 1.5. In nature, with mixed host types instead of the uniform fields of agriculture, nonspecific vertical resistance could be effective both in reducing the infection rate, and in being durable. Durability would be helped by reduced directional selection pressure towards virulence when pathogenic races spend much of their time on host plants not their "own"; and lack of specificity would make vertical resistance relevant even in endemic disease, provided that spores or other propagules disperse freely and widely. One concedes that nonspecific vertical resistance might be important and commoner than has generally been suspected; but numerical examples are difficult to formulate. Each situation would require its own particular formulation.

Chapter 7 Horizontal Resistance to Disease

7.1 Introduction

Horizontal resistance has been plagued by being given meaningful names that turn out in the end to have the wrong meaning. Plant pathologists have been slow to realize the value of nondescript appelations. Name your child Jack or Jane rather than Hercules or Rose because you never know how it will grow up. Every preconception embodied in synonyms for horizontal resistance has been found to be false. General resistance is a favorite, but, as we shall show, one of the features of horizontal resistance is great specificity. Polygenic resistance is commonly used, but it has yet to be shown that there is a single instance of resistance that is polygenic according to the accepted usage of the word, or that major variation in horizontal resistance involves more genes than in vertical resistance. Partial resistance has its advocates, but vertical resistance is also partial, and, as we shall discuss in the next chapter, some forms of horizontal resistance seem to be total and complete. Mature plant resistance or adult plant resistance, as a name used out of direct context with plant age is plainly silly, when seedlings also have horizontal resistance, and many examples are known of vertical resistance manifested only in the adult plant.

It is implicit in the definitions of vertical and horizontal resistance, as correlated and uncorrelated variation in host and pathogen, that genes for vertical resistance are relatively easy to identify and those for horizontal resistance difficult; this matter will be pursued later in the chapter. A gene for vertical resistance of comparatively trivial significance, such as the potato gene *R*4 for resistance to blight, is identified and, for that reason, has a substantial literature. Genes of major effect on horizontal resistance remain unidentified, and because they have no identity, have no literature. Inevitably, then, chapters in this book are long on vertical and short on horizontal resistance; and the relative paucity of literature of horizontal resistance might fail to bring home the importance of this form of resistance.

In Volume 3 of this series Robinson (1976) puts the case for horizontal resistance, and discusses how to use it. I have avoided repetition as much as is possible without breaking the continuity of the narrative of the topics discussed in this chapter and the next.

Both vertical and horizontal resistance have their merits and demerits, and it is no part of this book to try to inflate the one by deflating the other. One point should, however, be made, because it affects the emphasis of this and the next chapter. In practical agriculture, vertical resistance is reputed to have the advantage of relative simplicity in plant breeding. Horizontal resistance has the reputation of being unwieldy and complicated. This reputation is largely undeserved;

but in so far as it is deserved the objection can be removed only by understanding what horizontal resistance is about and how it is conditioned genetically and chemically. This understanding is what Chapters 7, 8, and 9 are aimed at.

7.2 Horizontal Resistance Not Disguised Vertical Resistance

Relatively so much is known about vertical resistance genes—*R* genes, *Sr* genes, *Lr* genes and the like—that some workers have tried to explain all resistance in terms of vertical resistance genes. That is, they regard horizontal resistance as some form of disguised or modified vertical resistance. The evidence they give does not survive critical analysis.

7.2.1 Hypothesis That Horizontal Resistance Is Buried Vertical Resistance

According to this hypothesis, when a gene for vertical resistance in the host is matched by the corresponding gene for virulence in the pathogen, compatibility is not fully restored; the resistance gene, even when buried by the matching virulence gene, still has a residual potency which (the hypothesis says) contributes to horizontal resistance. The evidence disputes this. It has neither been shown that such a residual effect, if it does indeed exist, would constitute horizontal resistance, nor has it been shown that the residual effect, if it exists, is more than entirely trivial in amount by comparison with ordinary horizontal resistance.

Martin and Ellingboe (1976) have stated the case for the buried resistance gene hypothesis. They worked with powdery mildew of wheat caused by *Erysiphe graminis tritici*. They used two near-isogenic lines of wheat, one with the gene for susceptibility *pm*4, the other with the gene for resistance *Pm*4. They inoculated these wheat lines with an avirulent isolate of *E. graminis tritici*, i.e., an isolate that could infect the *pm*4 type, but not the *Pm*4 type. They also inoculated the same wheat lines with three single-pustule isolates that were virulent on gene *Pm*4. They found that all four isolates (the avirulent isolate and the three virulent isolates) infected the susceptible *pm*4 line of wheat more efficiently than the three virulent isolates infected the resistant *Pm*4 line. Mutation to virulence in the pathogen did not fully compensate for mutation to resistance in the host; and Martin and Ellingboe suggested that an accumulation of virulent pathogen/resistant host combinations could constitute horizontal resistance.

Let us begin by an independent evaluation of the buried gene effect. The potato variety Vertifolia has two vertical resistance genes, *R*3 and *R*4, against *Phytophthora infestans*, i.e., when Vertifolia is attacked by blight it has two buried genes. The variety Capella has no *R* genes to bury. In an experimental plot observed by Kirste (1958), blight was first recorded on Vertifolia on August 22. By August 28, six days later, disease had increased to the stage at which several leaves per plant were infected. In contrast to this, Capella was first recorded blighted on

August 13. On September 20, 38 days later when observations ended, blight in Capella had still not yet reached the intensity found in Vertifolia on August 28. Capella is a very resistant variety, probably one of the most resistant to blight ever bred. The point of the comparison is that the resistance the buried genes *R* 3 and *R* 4 contributed to Vertifolia, if they contributed any, is absolutely trivial by comparison with resistance from other sources possessed by Capella.

The vertifolia effect, which we define as the loss of horizontal resistance in the process of breeding for vertical resistance, is discussed later. For the present we just note that the vertifolia effect and buried gene resistance are contradictory; there would be no vertifolia effect if buried genes contributed greatly to resistance.

We continue to assess buried gene resistance by observing that there are great variations of horizontal resistance even when there are no vertical resistance genes to bury. *Solanum tuberosum* sensu stricto, i.e., the South American species uncontaminated with Central American genes, has no *R* genes for resistance to blight. Within the species, however, there are very great variations in horizontal resistance, as Table 1.7 shows. This resistance owes nothing to buried *R* genes.

Other examples of horizontal resistance existing when there are no vertical resistance genes to bury are provided by the maceration diseases. Against them no vertical resistance is known, or, if the molecular hypothesis in Chapter 9 is correct, likely to be known. To give just two examples, grapevine cultivars differ widely in the resistance of their berries to post-harvest decay by *Botrytis cinerea*; and potato cultivars differ widely in the resistance of their tubers to *Erwinia carotovora* and *E. atroseptica*.

An objection of another sort is that all the evidence used to support the buried gene hypothesis comes from using particular isolates from the pathogenic population instead of using the population itself, without proving that the isolates represent a fit population. The three virulent isolates studied by Martin and Ellingboe differed very greatly in their ability to infect the *Pm* 4 type of wheat. One of these virulent isolates (the word virulent meaning no more than that the isolate could match the resistance gene) was able, after an initial short period of inferiority, to infect the *Pm* 4 type of wheat as efficiently as it or the avirulent isolate infected the *pm* 4 type. In contrast to this, another of the three virulent isolates was grossly inferior and had little efficiency at all times. In the field there would be strong selection for fitness in the pathogenic population; and experiments like Martin and Ellingboe's with highly variable isolates would carry conviction only if it were shown that the virulent isolates in the experiment were comparable with those fit to survive in the field.

The final objection is that the experiments have nothing to do with the difference between horizontal and vertical resistance. The experiments concern one locus, the *Pm* 4 locus, in the host and the corresponding one in the pathogen. Any demonstration that resistance is vertical or horizontal requires information from at least two loci in the host, and at least two in the pathogen (see Sect. 1.10, 3.26, and 8.3). Martin and Ellingboe's is just another example of confusion about the "quadratic check" on which they rely; and it needs to be repeated that the quadratic check, which concerns one locus in the host and one in the pathogen, is a check neither for the gene-for-gene hypothesis, nor for vertical or horizontal resistance (see Sect. 3.26 and 8.3).

7.2.2 Hypothesis That Horizontal Resistance Is Massed Vertical Resistance

This hypothesis has appeared from time to time in the literature. The most recent and explicit account was given by Parlevliet and Zadoks (1977). According to the hypothesis of massed vertical resistance, horizontal resistance can be explained as a polygenic resistance, where the individual genes are vertical and operating on a gene-for-gene basis with virulence genes in the pathogen.

Parlevliet and Zadoks base their conclusion on an analysis of variance in hypothetical data, unaware of the arithmetic fallacy involved when purely fictitious differential interactions appear in the analysis of variance although no differential interaction in reality exists (see Sect. 1.11 and the reference to Johnson and Taylor, 1976). Had Parlevliet and Zadoks used ranking methods, these would have neutralized the effect of the arbitrary changes of relative stretch in their example and left no evidence of a differential interaction.

The second objection is more general and basic. The massed gene hypothesis assumes that horizontal resistance is polygenic. We shall later in this chapter dispute that there is evidence that any resistance is polygenic; and, to argue more simply, there is evidence given in the next chapter that some horizontal resistance may even be monogenic. There is no evidence that more genes are substantially involved in horizontal resistance than in vertical resistance (see Sect. 7.5).

7.3 Examples of the Rapid Accumulation of Horizontal Resistance

As a preliminary to discussing gene numbers, five examples are considered here of the rapid accumulation of horizontal resistance under selection pressure, with the implication that relatively few genes are involved with great effect.

Simmonds and Malcolmson (1967) took clones of *Solanum tuberosum* from the Andes, which are the original home of *S. tuberosum*. The Andean potatoes were probably never ancestrally exposed to blight, *Phytophthora infestans* being indigenous to Central, not South, America. The clones used to start the experiment were all very susceptible to blight. They were subjected to mass selection in the field for resistance to blight, not directly but indirectly through selection for yield and conformity to modern horticultural type. Mass selection was continued through three sexual generations. Results, illustrated in Table 7.1, showed sharp gains of resistance as a result of selection over the first two generations, with no evidence for further gain during the third cycle of selection. This suggests that relatively few genes, mainly additive in effect, were involved.

Rapid gains in blight resistance through selection in potatoes of Andean origin have also been reported by Plaisted et al. (1975), unfortunately only in an abstract with few details.

Table 7.1. Increase in blight resistance in potatoes of Andean origin during mass selection over three sexual generations[a]

Generation	Percentage scoring 3 or less[b]	
	Greenhouse samples	Field samples
First	21	25
Second	45	78
Third	45	65

[a] Data of Simmonds and Malcolmson (1967).
[b] 1 = very resistant; 5 = very susceptible.

Destructive epidemics of leaf blight caused by *Helminthosporium turcicum* followed the introduction of hybrid maize in the United States. A program of breeding for resistance was started. Jenkins et al. (1954) used simple recurrent selection to increase resistance. There were nine source populations each of about 250 plants. The young plants were artificially inoculated to show symptoms before they flowered. In each population the ten most resistant plants were selected and interpollinated; pollen was collected from these ten plants, mixed in roughly equal proportions, and placed on the silks of the same ten plants. The seed so obtained was mixed, and used to grow plants to start the next cycle of selection. Within two cycles of recurrent selection, resistance was raised significantly within each of the nine population groups. In the third cycle, the improvement slackened its pace; the improvement in the third cycle was only one-third of what it had been in the second cycle. Jenkins et al. concluded that two cycles of recurrent selection were ample to bring susceptible populations to a level of resistance adequate for all purposes of practical farming. Hughes and Hooker (1971) also investigated *Helminthosporium* leaf blight in a study of gene action. They concluded from the general agreement between their results and those of Jenkins et al. that resistance to leaf blight is inherited as a comparatively simple character and conditioned by a small number of genes. They estimated the minimum number of effective factor pairs in the population as three to six. (These studies of Jenkins et al. and Hughes and Hooker did not involve the presumably distinct chlorotic lesion type of resistance monogenically conditioned by the *Ht*1 gene; see Sect. 3.26.)

Tropical rust of maize caused by *Puccinia polysora* is another example, differing only in that selection was natural over a continent and not man-guided in an experimental field. In 1949, or shortly before, *P. polysora*, a native of tropical America, crossed the Atlantic Ocean to Africa, and found maize there with its resistance dissipated after more than four centuries out of contact. An epidemic swept across tropical Africa, often killing plants before they could ripen. Unconscious and conscious selection, when farmers grew the most resistant lines they could (Cammack, 1961), quickly made the epidemic abate, and soon tropical leaf rust was a minor disease (Vanderplank, 1963; Robinson, 1976). Here too the evidence for swift improvement of resistance argues that relatively few genes were substantially involved.

Lupton and Johnson (1970) studied stripe (yellow) rust of wheat caused by *Puccinia striiformis*. The variety Little Joss is considerably resistant, and Nord Desprez very susceptible. They crossed these two varieties, and in the F2 generation began selecting for resistance. By the time the F4 generation was reached, i.e., after two cycles of selection, the horizontal resistance was considerable, and 84% of the population had horizontal resistance comparable with that of Little Joss. Of the single plants 40% had less than 0.1% of their leaf area attacked, 22% had between 0.1 and 0.25%, and 22% had between 0.25 and 1%. By comparison, Little Joss had 0.6% and Nord Desprez had 8.3%. At the time of the F5 generation, after three cycles of selection, 47% of the single plants had less than 1% of the leaf area attacked and 25% less than 4% compared with 0.7%, varying from 0.2 to 1.9%, for Little Joss and 48.5%, varying from 33.5 to 57.5%, for Nord Desprez. These disease assessments were made in plants exposed to natural epidemics, and the F5 generation was exposed to a more severe epidemic than the F4. Results for these two generations are therefore not strictly comparable. Nevertheless it is clear that a great amount of resistance was accumulated in only two cycles of selection, and results with *P. striiformis* follow the same pattern as those with *Phytophthora infestans* and *Helminthosporium turcicum*. There was, however, still segregation among disease readings in some lines of the F5 generation, whereas others showed a uniform level of resistance. This might indicate some genes of minor effect in addition to a few of major effect; it might also indicate that some parents chosen as resistant were in fact no more than escapes.

Sharp et al. (1976), also working with stripe rust of wheat, obtained results like those of Lupton and Johnson. Their program differed essentially from that of Lupton and Johnson in that they did not use a resistant wheat variety in their crosses. They made crosses between varieties of winter wheat that were susceptible or intermediate in their response to *P. striiformis*. The F1 generations were uniformly susceptible. Thereafter selection pressure was applied by using the five most resistant plants, out of 200 or more, to develop the next generation. In the F4 generation, after two cycles of selection, all plants were resistant or very resistant.

Kim and Brewbaker (1977) have made the most detailed analysis of all so far. They worked with leaf rust of maize caused by *Puccinia sorghi*, and set out to measure the genetic advance through selection. Their results in general were like those of Hooker (1967) with this disease; but their analysis was more detailed, and they worked in Hawaii in conditions of substantial disease. In crosses between moderately resistant and susceptible inbreds, the F2 population segregated continuously, with some plants as resistant as the parent inbreds, though seldom more resistant, and with little evidence of consistent transgressive segregation. By selecting the most resistant 5% of the plants, the genetic advance averaged 1.73 units (on a scale 1 = resistant, 7 = susceptible) per generation when the three most resistant inbreds had been used in the original crosses. Nonadditive variance was generally low. Consistent estimates indicated that the resistance of the most resistant inbred, Oh545, was conditioned by only two major, partially dominant genes.

7.4 Polygenic Inheritance

7.4.1 Confusion About Polygenes

Polygenic resistance is a term widely used in plant pathology, with varying meanings. Some of us have taken it to mean resistance conditioned by many genes, a meaning which would seem safe but, unless properly qualified, is in fact wrong. To many (if one may judge from the current literature) "polygenic" has come to mean "unidentified", and is a euphemism for "I do not know". Almost always, the term is used without proof that it is justified; and if editors had insisted on proof, the term would probably never have entered the literature of plant pathology.

This being so, it seems desirable to include a discussion of general matters concerning polygenes.

7.4.2 Discontinuous Variation and Qualitative Characters; Continuous Variation and Quantitative Characters

When Mendel's laws were rediscovered at the beginning of the century, differences in kind—i.e., qualitative characters—were sought and worked on. Laboratory animals, especially *Drosophila*, and plants were chosen that showed sharply discontinuous differences so that progenies could be separated into clearly different groups and counted to determine ratios such as 3:1. In this way the basic elements of genetics were confirmed or discovered: dominance, linkage, epistasis, structural inversions, etc.

We are concerned with terminology, and note that discontinuous variation and qualitative characters are for most practical purposes interchangeable.

It was realized, especially by those interested in the genetics of evolution, that continuous variation was equally important. There are no sharp segregations, and characters may exhibit any degree of expression between wide limits. For that reason they have been called quantitative characters; they differ in degree, but not in kind.

Here too there are usually no practical problems in terminology. Resistance can be called quantitative, if it varies continuously.

7.4.3 Polygenic Variation

Turn now from phenotypes to genotypes. Yule (1906), an early proponent of Mendelism, seems to have been the first to propose that continuous, quantitative variation could be conditioned by a large number of genes each individually having only a small effect. Mather (1943) was the first to use the word polygene. He made a detailed biometric study of polygenic inheritance in relation to natural selection. A large number of operative genes, each of equal effect, will condition continuous variation in the phenotype. The phenotypic frequency distribution will commonly be normal, but normality as distinct from continuity is not a necessary consequence. For example, nonallelic interactions could make the phenotypic frequency distribution skew even though it is determined polygenically.

Mather defined polygenic characters as characters controlled by many genes each having an effect small compared with the total relevant nonheritable fluctuations. There must not only be many genes, but all of them must have a small effect. Genes with major effects among those with small effects are excluded by definition. An extreme illustration shows why. Suppose resistance to some particular disease is conditioned by 20 genes; 20 presumably qualifies as "poly". Suppose further that 19 of them contribute a total of only 1% of the variance of heritable phenotypic segregation, the other gene contributing 99%. The resistance would be labeled monogenic; indeed, with inevitable nonheritable variation present, there would be no way of showing that it was not monogenic. So too with less extreme illustrations it can be seen that for inheritance to be called polygenic there should be no genes of major effect involved.

7.4.4 The Untrue Converse

Polygenic inheritance means that the variation of the phenotype will be continuous. The converse is entirely untrue. Continuous phenotypic variation does not necessarily mean control by many genes. Mather's studies and those of most others were about genes of equal, or at least similar, effect. As soon as the postulate of similar effects is removed, continuous phenotypic variation ceases to necessarily mean that many genes are involved.

Thompson (1975) demonstrated a theoretical model of a continuous normal phenotypic distribution determined by the segregation of only three genes, one of these having three times, and one twice, the effect of the third gene. The distribution approached a normal distribution, even though only two loci accounted for more than 90% of the variance. If nonheritable variation is large, even fewer genes might be enough to give continuous variation. When heritability is about 50%, samples from an F2 might vary almost continuously even though only one locus were segregating. Clearly, there is no justification for automatically assuming that a continuous distribution requires the segregation of a great number of polygenes.

The proposition, that continuous phenotypic variation does not necessarily indicate polygenic inheritance, is in practice more important than the original proposition, that polygenic inheritance conditions continuous variation, because it is the phenotypes we see and evaluate. One is easily trapped into incautious terminology. Continuous or discontinuous variation, quantitative or qualitative characters, and horizontal or vertical resistance are terms pertaining to the phenotype. Polygenic is an adjective for a particular sort of genotype, clearly circumscribed. Yet there is frequent talk in the literature of polygenic resistance when it is only known that the resistance is quantitative. In plant pathology, polygenic is in danger of fast becoming an adjective for a phenotype. Thus proceeds the degradation of vocabulary.

7.5 The Number of Genes Conditioning Horizontal Resistance

Continuous variation of segregates of the sort commonly associated with horizontal resistance can, in Thompson's model, be conditioned by as few as three genes, or less if nonheritable variation is considerable. How many genes, then, are substantially involved in the variation of horizontal resistance? The answer seems to be, one or more.

Monogenic horizontal resistance is a topic of the next chapter. In Section 7.3 evidence was given to suggest that much horizontal resistance was conditioned by relatively few genes, these genes having major effects. The diseases were Phytophthora blight of potatoes, leaf blight of maize caused by *Helminthosporium turcicum*, the leaf rusts of maize daused by *Puccinia polysora* and *P. sorghi*, and stripe rust of wheat caused by *P. striiformis.*

There is evidence from other diseases that much of the variance in segregating populations is conditioned by relatively few genes. Parlevliet (1976a) concluded that horizontal resistance in barley to *Puccinia hordei* is conditioned by one recessive gene with a fairly large effect and four or five minor genes with additive effect (see Sect. 7.6). In other experiments Parlevliet and Kuiper (1977) concluded that horizontal resistance in barley to *P. hordei* is governed by more, but not many more, than three genes. Grümmer et al. (1969) concluded that horizontal resistance to *Phytophthora infestans* in the tomato cultivar Atom is controlled mainly by two incompletely dominant genes, with minor modifying genes. Lim (1975) studied the reaction of eight maize inbreds to *Helminthosporium maydis* race T. Genetic resistance (i.e., resistance conditioned by nuclear genes) varied considerably in inbreds with Tms cytoplasm, and Lim concluded from his analyses that resistance was largely dominant and controlled by only a few genes. In this and all the other examples which have been discussed it is assumed that linkage is of little or no importance.

About the upper limit to the number of conditioning genes there is no evidence. Two general points about gene numbers involved in horizontal resistance merit mention.

First, there is no evidence now available to suggest that more genes are substantially involved in variations of horizontal resistance than could potentially be involved in variations of vertical resistance. There are probably 20 or more *Sr* and *Lr* genes for vertical resistance in wheat. Are the numbers greater for horizontal resistance? Many standard wheat cultivars have three or more vertical *Sr* genes; Selkirk has five: *Sr*6, *Sr7b*, *Sr9d*, *Sr*17 and *Sr*23, and Manitou five: *Sr*5, *Sr*6, *Sr7a*, *Sr*12, and *Sr*16 (Green, 1976b). One isolate of *P. graminis tritici* is known, 151-QSH (Roelfs and McVey, 1975), that is virulent on 18 *Sr* genes; they are *Sr* 5, 6, 7*a*, 8, 9*b*, 9*d*, 10, 11, 12, 15, 16, 17, 18, 19, 20, 21, 23, and 28. A wheat plant with all these genes is not yet known, but presumably, despite them all, would succumb to race 151-QSH if they met; and the resistance would be vertical. Against this, set the horizontal resistance of maize inbred Oh 545 to *Puccinia sorghi* that is conditioned by two gene pairs, or the horizontal resistance in maize to *P. polysora*, *Helminthosporium turcicum*, *H. maydis*, in wheat to *Puccinia strii-*

formis, in barley to *P. hordei*, and in potatoes and tomatoes to *Phytophthora infestans*. The estimates of gene pairs involved in the horizontal resistance run from two to six, and are of much the same order as the five *Sr* genes in Selkirk or Manitou.

Second, as an obvious corollary, there is no evidence that the number of resistance genes determines why horizontal resistance is usually stable and vertical resistance not. Number per se does not determine stability (Vanderplank, 1975). To repeat an example relevant in the present context, Selkirk wheat with five genes for vertical resistance has been severely attacked by stem rust in the south of North America, whereas maize inbred Oh 545 with two genes for horizontal resistance is little attacked by *P. sorghi* even in Hawaii where leaf rust is prevalent.

7.6 The Anonymity of Horizontal Resistance Genes

The genes conditioning vertical resistance in a plant can usually be directly identified; those conditioning horizontal resistance cannot (if monogenic and perhaps digenic resistance are excluded). This in itself tells a significant story.

Fortuitously, two papers by Parlevliet (1976a, b) were juxtaposed in publication. They were about leaf rust of barley caused by *Puccinia hordei*, and illustrate the point. One paper referred to seven genes for vertical resistance. All have been identified: *Pa*, *Pa*2, *Pa*3, *Pa*4, *Pa*5, *Pa*6, and *Pa*7. The other paper concludes that in the cultivars Vada and Minerva horizontal resistance (as manifested in the latent period) is conditioned by one recessive gene with a fairly large effect, and four or five minor genes with additive effect. The F 3 data corroborate this, and suggest the action of at most six genes, assuming no linkage. None of these genes was identified individually.

The number of genes discussed in Parlevliet's two papers is roughly the same. The difference is in the identification.

One identifies vertical resistance genes, in the sense that genes *Pa*, *Pa*2 ... *Sr*5, *Sr*6, ... *R*1, *R*2, ... are identified, by testing the host plant against appropriate isolates of the pathogen, i.e., by using differential races. Thus it is that one can say with confidence that the gene *R*1 in a modern potato cultivar is the same gene as one found in the Mexican wild species, *Solanum demissum*. So, too, by using appropriate isolates of *Puccinia graminis tritici* one could, with enough effort, determine which lines in the International Wheat Rust Nursery carry, say, the gene *Sr*8. The identification is possible because of differential races, which necessarily imply correlated variation in host and pathogen; and correlated variation is the definition of vertical resistance.

No such identification is possible for horizontal resistance genes, because, by definition, there is no correlated variation. True, one can at times identify the chromosome on which a horizontal resistance gene occurs, or associate the gene by close linkage to some other, easily identified gene. If in this way we identify a horizontal resistance gene, the identification is indirect; it is the chromosome or linked gene that is directly identified, not the horizontal resistance gene.

Correlated variation in host and pathogen, the basis of vertical resistance, allows vertical resistance genes to be identified by differential races of the pathogen. Uncorrelated variation in the host and pathogen, the basis of horizontal resistance, leaves horizontal resistance genes unidentified, except indirectly through the identification of a chromosome or a linked gene. The absence of directly identified horizontal resistance genes (except in monogenic and perhaps digenic resistance) is explained by horizontal resistance's definition as uncorrelated variation and, seen from the opposite angle, vindicates the definition.

Monogenic resistance can be identified but falls outside the story. Gene identification is the process of telling genes apart; and if there is only one gene there is no process of identification, except for the presence or absence of the resistance allele.

7.7 Specificity in Horizontal Resistance

In many, perhaps most, diseases horizontal resistance is largely conditioned by genes at relatively few loci, and, associated with this, is largely specific.

Potato cultivars can be highly resistant, horizontally, to *Phytophthora infestans*, but very susceptible to *Alternaria solani*. Conversely, some cultivars with considerable resistance to *A. solani* are very susceptible to *P. infestans*. There is no mutual exclusion; some potato cultivars have moderately good horizontal resistance to both pathogens, and some have great susceptibility to both. Evidently there are at least two systems of horizontal resistance at work, each specific for its disease. So too one finds maize lines with substantial horizontal resistance to rust caused by *P. sorghi*, but great susceptibility to *Helminthosphorium turcicum*, and vice versa. Similarly, maize lines can have substantial horizontal resistance to *P. sorghi*, but great susceptibility to *P. polysora*. Here resistance is specific, both in the common meaning of the adjective, and in its botanical meaning.

Specificity makes one doubt whether the contribution to horizontal resistance of a general or generalized system of defense controlled by many genes is large. If such generalized features as an open canopy, a thick or water-repellent cuticle, thick cell walls to be penetrated, or low (or high) sugar content made important contributions, one would have expected the same system of horizontal resistance to be active against many diseases. It is not disputed that such shared resistance may occur; it is simply doubted whether this resistance is of great magnitude in many diseases. Admittedly, much of the evidence is inconclusive, and can be interpreted in two ways. Field maize usually has substantial horizontal resistance to all the important local pathogens. Parallel selection may account for much of this, but there is no evidence to prove that it accounts for all of it.

Let us return to *P. sorghi* and *P. polysora*. Maize in Africa has no recorded history of great epidemics of *P. sorghi*; it has substantial horizontal resistance, and evidently this was on the whole adequate. However, when *P. polysora* came to Africa in 1949 or soon before (and the evidence of herbarium sheets shows conclusively it was not present long before 1949), maize was shown to be intensely susceptible to it. The resistance to *P. sorghi* was evidently highly specific. Maize was horizontally resistant to *P. sorghi*, but susceptible to *P. polysora*. *P. sorghi* was unable to exploit whatever process it was in maize that made it very susceptible to *P. polysora*. Why not? *P. sorghi* infected maize with the normal susceptible type "3" or "4" pustules. Indeed, the pustules were of much the same type as those produced by *P. polysora* when it arrived, and to the uninitiated the two rusts were not easily distinguished in the field. The climate was conducive to severe infection by *P. sorghi*, as shown by the heavy disease often noted in susceptible (and therefore inevitably discarded) inbred lines and simple crosses that commonly appear in maize breeders' plots. The only answer seems to be that in some details the processes that allowed great susceptibility to *P. polysora* could not be exploited by *P. sorghi*. *P. sorghi* was specifically barred, and on the evidence the barring was principally by only two gene pairs.

7.8 The Vertifolia Effect

The vertifolia effect has been defined as the loss of horizontal resistance in the process of plant breeding for vertical resistance (Vanderplank, 1963). The implications are disturbing. Plant breeders concentrating on vertical resistance may often have allowed the level of horizontal resistance to retrogress. The further implication is that increased horizontal resistance may decrease fitness, when fitness is measured in the absence of disease.

The original evidence came from the progress of blight caused by *Phytophthora infestans* in plots of potato varieties with or without *R* genes. In Kirste's (1958) data there were 12 late-maturing varieties appropriate for analysis, six with and six without *R* genes. The evidence concerned the time taken for blight to progress from rating 1 (a very mild attack) to rating 3 (medium infection). In the six varieties with *R* genes the average date for rating 1 was August 15 and for rating 3 August 30, a span of 15 days. In the six varieties without *R* genes the corresponding dates were August 7 and September 2, a span of 26 days. The start of the blight epidemic was delayed by *R* genes (see Sect. 6.17), but, once started, was much faster when *R* genes were present. The infection rate *r* was in fact 26/15 times faster in varieties with *R* genes than in those without. Weather conditions contributed nothing to the faster infection, because the relevant dates for the progress of blight when there were no *R* genes straddle those when there were. All the weather that affected blight in varieties with *R* genes between August 15 and August 30 was weather that affected blight in varieties without *R* genes during the same period.

Five of the six varieties have the gene *R* 1. Vertifolia has the genes *R* 3 and *R* 4. It had a particularly high infection rate. In this cultivar blight rating 1 was reached late, on August 22, reflecting the fact, established by independent surveys, that combined virulence on *R* 3 and *R* 4 was very rare in the population of *P. infestans*; but blight rating 3 was reached on August 29, six days later. The infection rate *r* was 26/6 times faster in Vertifolia than in varieties without *R* genes.

Other data concerning potato blight confirm that the infection rate is higher when *R* genes are present. The cultivars which Schick et al. (1958) rated on field performance as very resistant took 16 days for blight to progress from rating 1 to rating 3 when *R* genes were present, but 32 days when they were absent. With those rated as moderately resistant the corresponding figures were 12 and 25 days. Resistance as assessed by Schick et al. was a mixture of vertical, which delays the start of an epidemic, and horizontal, which slows the epidemic down after it has started. If one separates out the horizontal resistance and measures it as being inversely proportional to the time taken for blight to progress from ratings 1 to 3, the infection rate r was about twice as fast in varieties with *R* genes as in those without. Data from the Netherlands (Hogen Esch and Zingstra, 1957) point in the same way.

Lupton and Johnson (1970) have documented a vertifolia effect in relation to *Puccinia striiformis* in wheat, and Robinson (1976) has discussed the effect at length.

7.9 Resistance and Loss of Fitness

Plants bred in the absence of a disease have, or tend to have, lowered horizontal resistance. This is the evidence of the vertifolia effect: plant breeders making their selections among host populations protected from the disease by vertical resistance tend unwittingly to select progeny with low horizontal resistance. By implication, optimal fitness in a population not threatened by disease does not coincide with optimal fitness in a threatened population.

This is also the evidence of many great epidemics. Populations of plants that have grown up for generations unthreatened by a pathogen often succumb badly to the pathogen when they meet it for the first time or after a long period of separation. Examples are diverse. Among virus diseases one thinks of swollen shoot disease when cacao from America met African viruses in Africa, and of potatoes in central Europe attacked by potato virus Y^n (tobacco veinal necrosis virus) when it was introduced. Among bacterial diseases, the pear tree from the Old World met the fireblight organism *(Erwinia amylovora)* in the New, and suffered badly. Among fungus diseases the list of examples is long. At about the same time, *Phytophthora infestans* in potatoes and *Uncinula necator* in grapevines crossed the Atlantic to devastate fields and vineyards in Europe, where they had previously been unknown. Other fungus pathogens new to Europe that started destructive epidemics were *Plasmopara viticola*, causing downy mildew in grapevines, *Peronospora tabacina*, causing blue mold in tobacco, *Pseudoperonospora humuli*, causing downy mildew in hops, and *Sphaerotheca mors-uvae*, causing powdery mildew in gooseberries. Among pathogens new to America and starting destructive epidemics there were *Cronartium ribicola*, causing blister rust in white pine, *Ceratocystis ulmi*, causing Dutch elm disease, and *Endothia parasitica* that practically destroyed the American chestnut. *Puccinia polysora* was new to Africa when it swept across the continent in an epidemic.

A frequent feature of these epidemics, particularly among annuals or clonal host plants, was how quickly the epidemic abated, largely due to the new selection pressure of the pathogen. The history of potato virus Y^n is recent and well documented. When it spread through the potato fields of central Europe in the late 1950s, there was a sufficient diversity of potato cultivars and advanced lines in breeders' fields to absorb the shock within a few years; favorite old cultivars like Ackersegen were replaced by others much less susceptible, and, although spread of the virus cannot easily be controlled by insecticides, the potato industry has suffered little harm in the long term. In populations of long-lived forest trees genetic change is much more difficult, and with chestnut blight in America the epidemic ran its course to the virtual extinction of the host.

One concludes from this that there is a price to pay, in lost fitness, for horizontal resistance. Of course, resistance itself is a gain of fitness in another direction, and a new balance of optimal fitness in the environment of disease is presumably struck. We cannot assess the loss of fitness of resistant hosts relative to susceptible hosts in the absence of the pathogen. Gene recombination may well reduce the loss. Vanderplank (1963) suggested that the increased resistance to *Puccinia polysora* quickly developed by open-pollinated maize in Africa could have been due to an accumulation of homozygotes, in which case gene recombination might in the course of time have reduced the loss of heterosis.

Advocates of vertical resistance have no cause to feel smug about this apparent disability of horizontal resistance. There is equal evidence that vertical resistance, too, may bring with it a loss of fitness when disease is absent. In the absence of a pathogen and selection pressure from it, alleles for vertical susceptibility are common, and those for resistance rare in host populations, which indicates a fitness advantage in susceptibility. The matter was discussed in Section 2.5, where the purpose was to show that what we call genes for susceptibility have some function in the plant other than to welcome the parasite. The evidence, appropriately re-read, is more precisely for the superiority, and not just equality, of alleles for susceptibility over those for resistance in the absence of the pathogen.

We can only conclude that plant breeders should not try to introduce more resistance, either horizontal or vertical, than is required. Very high resistance may seem desirable, but we may sometimes pay too high a price in terms of fitness. Plant breeders left to themselves are unlikely to fall into this error, because their attention is primarily on yield and quality, but plant pathologists can at times become too enthusiastic about disease resistance.

Multilines, grown from a mixture of seed of lines each differing by a resistance gene, provide a counter to excessive loss of fitness as a result of increased resistance when this resistance is vertical, but not when it is horizontal.

7.10 Epidemiological Effects of Horizontal Resistance

The effect of resistance on disease in populations of plants, which is what epidemiology is about, is as much a part of our subject as is the effect on disease in individual plants.

7.10.1 Effect on the Infection Rate

Horizontal resistance reduces the infection rate. (So too does nonspecific vertical resistance.) "Slow rusting", or corresponding terms for other types of disease, is sometimes used as a synonym for horizontal resistance; and it is pure tautology to say that disease in a slow-rusting cultivar has a low infection rate.

Four factors of resistance condition a low infection rate: a lower proportion of spores (or other units of infection) that manage to initiate a lesion, a lower average spore production by a lesion, a shorter period of spore production by the lesion, i.e., a shorter period of infectiousness, and a longer period of latency. Examples of all four factors have appeared largely in the literature over the years; often there is a bias in experimental results, as when an artificially heavy dose of inoculum puts special emphasis on the first factor. What is missing from the literature is much evidence about how far the factors are independent or interdependent genetically and chemically. Are there genetic or molecular curbs common to, say, resistance to infection by spores on the one hand and reduced spore production on the other? Parallel selection pressures might change all four factors simultaneously—the potato cultivars Capella and Pimpernel have resistance to blight operated by all four factors—but this is no real evidence for common genetic or chemical action. However, it seems plausible to suppose a priori that resistance to spread of mycelium through tissues, and hence genetic and chemical factors that condition this resistance, could affect three of the factors: the rate of spore production in a lesion, the duration of spore production, and the period of latency.

7.10.2 Delay of the Start of an Epidemic

Resistance to infection, in the form of a reduced proportion of spores that manage to initiate lesions, would delay the start of an epidemic, as measured by the date at which some particular level of disease is noticed. If resistance reduces the proportion of spores that initiate lesions from, say, 10% to 1%, the epidemic would be delayed by the period required for disease to increase ten-fold.

Delays of this sort are not likely to be as substantial with horizontal as with vertical resistance.

7.10.3 Contraction of the Scale of Spread of an Epidemic

It is a feature of epidemic disease that it spreads. Indeed, this feature is inherent in the definition of epidemic disease as sporadic disease. (Sporadic does not mean relative harmlessness!) *Puccinia graminis* spreads from overwintering sites, through winter wheat, to summer wheat fields. *Phytophthora infestans* spreads from infected plants in diseased cull piles to plants in adjacent fields, and from them across the countryside. A well-developed epidemic advancing over a large area has an advancing front, with high levels of disease behind the front and low levels ahead where the epidemic has recently advanced into healthy fields. Let us define isopaths as lines of equal disease. Then, following the reasoning of Vanderplank (1975), the distance between any two isopath lines (e.g., the distance between the line for 10% disease and the line for 1% disease) is determined by the infection rate r, other things being equal. Put differently, the gradient of disease at the epidemic's front is determined by the infection rate. The greater the horizontal resistance, the lower the infection rate, the steeper the gradient, and the more compact the epidemic's front will be. This is so, even if resistance has no effect on the gradient of deposition of spores from a point source. (If resistance does not affect the growth form of the plant and the canopy in the field, there seems to be no reason why it should affect the average pattern of flight of a spore from a point source.)

MacKenzie (1976) suggested using the spread of disease from a point source into suitably arranged plots as a means of identifying resistance to stem rust in wheat lines. The distances and the durations of time involved in his experiments were short, and it was primarily the gradient from a point source that was measured. The method has promise for the purpose MacKenzie had in mind, but it must greatly underestimate the effect of horizontal resistance on spread of disease in a developed epidemic.

Chapter 8 Selective Pathotoxins in Host–Pathogen Specificity

8.1 Introduction

The terms selective pathotoxins and host-specific toxins are used for compounds that affect only those plants within a species that are susceptible to the fungus or other organism that produces them.

Only a few diseases have been shown to be produced by selective pathotoxins (although the real number may be great), and the best-known examples are listed in Table 8.1.

Many of the known selective pathotoxins are produced by members of the genus *Helminthosporium*. They seem, however, to be rather diverse. On present evidence, some are oligopeptides, some are terpenoid, and one is a glycoside. Nevertheless the known selective pathotoxins share common features. They have been isolated, at least in a state of partial purity, and when inoculated into the host plant have proved to be substitutes for the pathogens themselves in producing symptoms. Indeed, in many plant breeding programs the pathotoxin can be used instead of the pathogen to screen out the susceptible plants in a segregating population. The host–pathogen specificity is great, and, relative to most other diseases, more easily followed in laboratory experiments. For this reason, it behoves us to investigate examples in some detail.

The number of known examples of selective pathotoxin diseases, though small, is enough to provide important evidence. The evidence is for horizontal resistance that is specific and, commonly, monogenic; and for horizontal resistance which the pathogen can overcome only when its pathogenicity exceeds a threshold level.

Table 8.1. Some diseases caused by selective pathotoxins[a]

Disease	Host	Pathogen
Victoria blight	Oats	*Helminthosporium victoriae*
Southern leaf blight	Maize	*H. maydis*, race T
Leaf spot	Maize	*H. carbonum*
Eye spot	Sugarcane	*H. sacchari*
Yellow leaf blight	Maize	*Phyllosticta maydis*
Milo disease	Sorghum	*Periconia circinata*

[a] Literature is cited by Scheffer and Yoder (1972) and Wheeler (1975). References specially relevant to this chapter are given in the text.

8.2 Some Diseases Caused by Selective Pathotoxins

8.2.1 Victoria Blight of Oats

Victoria, a South American oat cultivar, was brought to North America and used extensively in plant breeding because it is resistant to many races of crown rust *(Puccinia coronata avenae)* and smut *(Ustilago avenae)*. Cultivars derived from crosses with Victoria were released in Iowa in 1942, and by 1945 were used on 98% of the oat acreage in Iowa and 50% in the United States as a whole (Pringle and Scheffer, 1964). These cultivars met *Helminthosporium victoriae*, as it was afterwards called, apparently existing as a saprophyte or weak parasite on grasses. In 1944 *H. victoriae* was first seen on oats, and in 1946 it was first described, and the disease it caused named Victoria blight. Epidemic outbreaks soon caused Victoria to be replaced by Bond and other cultivars without Victoria ancestry.

H. victoriae is soil- and seed-borne, causing basal stem and root necrosis, together with a leaf blight. Meehan and Murphy (1947) applied cell-free filtrates from cultures of the fungus to roots of intact seedlings. Seedlings of Victoria oats were killed, even in filtrates diluted 90 times with water, whereas seedlings of resistant varieties were unharmed. Since 1947 a large literature has come into being, to confirm Meehan and Murphy's work and to develop the concept of selective pathotoxins.

The toxin is highly potent. Luke and Wheeler (1955) reported culture filtrates from a potent isolate of *H. victoriae* which caused 50% reduction of root growth in susceptible oats, even at a dilution of one in 1 200 000. Resistant varieties of oats were unaffected except by filtrates about 400 000 times as concentrated as those toxic to susceptible varieties (Wheeler, 1975). Rye, wheat, barley, rice, maize, lettuce, tomato, pea, beet, carrot, cucumber, okra, radish, mustard, bean, and spinach are insensitive to the toxin (Luke and Wheeler, 1955).

When *H. victoriae* invades susceptible host tissue, it at first causes only a slight clearing of cells in advance of the hyphae. About two to three days after inoculation, host cells start to collapse, presumably because of the toxin, and damage spreads from the site of infection. Susceptibility is controlled by a single dominant gene, although modifying factors may be involved. Resistance in oats to *H. victoriae* and to its toxin are the same genetically, and seedlings may be screened in breeding programs by exposing them to the toxin rather than to the fungus.

The toxin, victorin, is unstable and its structure is not yet completely known. The molecular mass is relatively low, between 800 and 2000. The molecule has an oligopeptide fragment containing five or six amino acid residues linked to a sequiterpene (Pringle, 1972). Because of the instability of the toxin, it has not been isolated in a pure state and it is not known with certainty how it exerts its effect. Summarising evidence from all sources, Scheffer and Yoder (1972) suggest the hypothesis that an initial lesion occurs after exposure to toxin, that the lesion may be in the plasma membrane, and that a protein may be involved as a receptor substance. They further suggest that, if it is accepted that susceptible cells have a receptor substance or site, it is logical to assume that resistant cells lack this substance or site. In later comments, however, Scheffer (1976) stated that toxin was detected in the protein fractions of both resistant and susceptible plants, and that membrane preparations from resistant and susceptible plants bound small but equal amounts of toxin. This indicates, Scheffer believes, that victorin is not firmly bound to a specific component of susceptible cells. It is, however, possible that the part of the molecule involved in binding is not the part that reacts toxically with susceptible but not with resistant varieties. It need not be.

H. victoriae is known in its sexual stage *Cochliobolus victoriae* (Nelson, 1960), and crosses with avirulent isolates have been used to show that virulence is monogenetically controlled.

8.2.2 Southern Maize Leaf Blight

This disease caused a severe epidemic in maize in the southeast of the United States in 1970, and resulted from using cytoplasmic male sterility as an aid to producing hybrid maize seed. The cytoplasm concerned, Texas male sterile (Tms) cytoplasm, made the plant susceptible to *Helminthosporium maydis* race T. The immediate cause that triggered off the epidemic has not been properly established. The weather favored the disease, but not to an unusual extent.

The change from *H. maydis* race 0 to *H. maydis* race T brought about a change of symptoms. The common symptom with race 0 is a leaf blight or narrow spot from 1 to 2 cm long. Race T on Tms cytoplasm produces lesions on the leaf sheath and husk and a gray to black ear rot, in addition to leaf lesions; and the leaf lesions are longer and usually have a chlorotic border.

There is still some doubt about the toxins of *H. maydis* race T. Karr et al. (1974), using a leaf bioassay test, isolated four terpenoid, host-specific toxins from cultures of *H. maydis*, but more work is needed to understand the toxic effect as a whole.

8.2.3 Eye Spot Disease of Sugarcane

The disease caused by *Helminthosporium sacchari* in sugarcane has been the pathotoxin disease most studied in recent years. The toxin, named helminthosporoside, was isolated by Steiner and Strobel (1971). Like the other selective pathotoxins produced by *Helminthosporium* spp. it has a low molecular weight, but unlike the others, or most of them, it contains no peptide moiety; indeed, it contains no nitrogen at all. Although another toxin may also be present, the action of helminthosporoside correlates with that of *H. sacchari* itself. Purified helminthosporoside had the same biological specificity to sugarcane clones as *H. sacchari* itself; in tests with 11 sugarcane clones, inoculation with helminthosporoside produced characteristic runner lesions symptoms in the six susceptible, but none in the five resistant clones, just as inoculation with *H. sacchari* itself did.

Helminthosporoside is produced both in artificial culture of *H. sacchari* and in lesions in susceptible sugarcane clones.

Helminthosporoside is an α-galactoside of low molecular weight. It is the galactose residue of the toxin that is bound to the host during pathogenesis, and the binding is specific for α-galactosides. The disaccharide melibiose and the trisaccharide raffinose, both of them α-galactosides, are also bound, but lactose, a

β-galactoside, is not. Strobel (1973a, 1975) demonstrated this neatly. Pieces of sugarcane leaf were split down the middle. Half of the leaf was placed in a solution of melibiose or raffinose, and the other half in water or a solution of lactose. A day later both halves were injected with the toxin. Symptom development was inhibited in the half leaves that had taken up the melibiose or raffinose, but not in the half leaves that had been in water or a lactose solution. Also, in tests with a membrane preparation from a susceptible sugarcane clone, melibiose, raffinose and methyl-α-galactopyranoside reduced the ability of helminthosporoside to bind with the membrane preparation, whereas lactose, glucose, sucrose, galactose itself, and methyl-β-galactopyranoside did not. Conversely, cells from resistant sugarcane clones do not actively take up raffinose (Strobel, 1974).

Strobel and Hess (1974) reported that helminthosporoside is bound to the plasma membrane of leaf cells of susceptible sugarcane clones. This result was confirmed independently by Thom et al. (1975), in their work on pure membrane preparations obtained from sugarcane cells grown in suspension culture. The binding is by a protein on the plasma membrane (Strobel and Hess, 1974). Only the protein from susceptible sugarcane clones binds helminthosporoside; that from resistant clones does not (Strobel, 1973b). Strobel (1973a) correlated to the 0.05 level of significance the relationship between the reaction of 29 clones of sugarcane to helminthosporoside, and the ability of a membrane preparation from any given clone to bind the toxin.

Leaves of susceptible clones pretreated with detergent (Triton X-100) became resistant to the effect of helminthosporoside, presumably because the binding protein is removed from the membrane by the pretreatment (Strobel, 1973a).

Strobel and Hess (1974) prepared a rabbit antiserum to the purified binding protein. Leaves of susceptible sugarcane clones were treated with this antiserum, which made them resistant to the effect of inoculation with helminthosporoside. Treatment with control serum, without the specific antibodies, had no protective effect. They also tested the agglutination of isolated wall-less protoplasts with antiserum to the binding protein. Protoplasts (from susceptible sugarcane clones) agglutinated in the presence of antiserum to the purified binding protein, but not in the presence of control serum. These tests, with others, suggest that the binding protein is in the outer membrane of the cell.

Strobel and Hapner (1975) were able to transfer susceptibility from susceptible sugarcane clones to leaf cell protoplasts of resistant sugarcane and of tobacco. Toxin-binding protein from membranes of a susceptible sugarcane clone was adsorbed on to wall-less protoplasts from resistant sugarcane and tobacco. These protoplasts then became susceptible to helminthosporoside, and also took up raffinose.

Strobel (1973a, b) estimated by gel electrophoresis that the toxin-binding protein has a molecular weight of about 49000; by gel filtration the estimate was 45000. The protein is a tetramer of four similar subunits. The proteins from susceptible and resistant sugarcane clones are similar; analyses so far reported show slight differences in the content of glycine, lysine, serine, and glutamic acid. Enough pure binding protein is available, and amino acid sequences are now being investigated (Strobel, personal communication 1977). This will presumably pinpoint the difference between the proteins from susceptible and resistant clones.

8.3 Selective Pathotoxins Unconnected with Gene-for-Gene Systems

Victoria blight of oats and other selective pathotoxin diseases are commonly confused with gene-for-gene diseases, the confusion arising from the mistaken "quadratic check". The quadratic check concerns itself with alleles at one locus in the host and alleles at one locus in the pathogen. Flor's gene-for-gene hypothesis requires information about at least two loci in the host, and at least two in the pathogen.

Flor (1956, 1959) clearly defined his hypothesis. For each gene conditioning resistance in the host there is a specific gene conditioning pathogenicity in the parasite. In an alternative form, for each gene conditioning resistance in the host there is a complementary or reciprocal gene conditioning pathogenicity in the parasite. The words "specific", "complementary", and "reciprocal" are Flor's. In passing, we notice that Flor writes of "each" resistance gene, which clearly indicates that he had in mind more than one; but let us examine basic alternatives.

Isolates of *Helminthosporium victoriae* have been found that lack pathogenicity. They do not attack even Victoria oats. Normal isolates attack Victoria and related oat varieties, but not Bond (or Landhaufer or Santa Fe or other varieties at present resistant). Table 8.2 lists these two sorts of isolates, the nonpathogenic and the normal. Further, it adds a hypothetical situation in which a new isolate of *H. victoriae* is found. In Case A of Table 8.2 the new hypothetical isolate attacks Bond but not Victoria; the new hypothetical isolate differs qualitatively from the known. In the Case B it attacks both Bond and Victoria. It differs quantitatively.

Case A exemplifies a gene-for-gene system, and its position of Table 8.2 corresponds to Table 3.8. In general terms, race 1 of a pathogen attacks $R1$-types of the

Table 8.2. The reaction of two oat varieties to three isolates of *Helminthosporium victoriae*, two actual, the other hypothetical

(A) Data showing hypothetical qualitative resistance

Isolate	Variety	
	Victoria	Bond[a]
Not pathogenic	Resistant	Resistant
Normal	Susceptible	Resistant
Unknown, hypothetical	Resistant	Susceptible

(B) Data showing hypothetical quantitative resistance

Isolate	Variety	
	Victoria	Bond
Not pathogenic	Resistant	Resistant
Normal	Susceptible	Resistant
Unknown, hypothetical	Susceptible	Susceptible

[a] Bond or other variety at present resistant.

host, but not $R2$-types; race 2 attacks $R2$-types but not $R1$-types. This meets the minimum requirement that for every gene for resistance in the host there must be a "specific", "complementary" or "reciprocal" gene for virulence in the pathogen.

Case B gives no evidence for specific, complementary, or reciprocal genes. The hypothetical isolate attacks both Victoria and Bond, unspecifically. (If it is assumed that the hypothetical isolate has two genes for virulence, one specific for Victoria and one for Bond, the onus would be on the experimenter, by appropriate crossing with an avirulent isolate, to produce a segregate that attacks Bond but not Victoria.)

Until such time as anyone finds an isolate of *H. victoriae* that attacks Bond or other currently resistant variety *but does not attack Victoria*, i.e., until such time as Case A becomes real, we can ignore the notion that Victoria blight of oats is a gene-for-gene disease. The same comment holds for other diseases of this group: in no disease caused by a selective pathotoxin has specificity sensu Flor been demonstrated.

8.4 Evidence for Horizontal Resistance

In establishing host–pathogen relations, the toxins have an advantage over the pathogen itself. Consider *Helminthosporium maydis* on maize. The toxin from cultures of *H. maydis* race T is much more toxic to plants with Tms cytoplasm than to those with normal cytoplasm. Nevertheless there are toxin concentrations high enough to produce symptoms even on plants with normal cytoplasm, and, at the other extreme, concentrations too low to affect even plants with Tms cytoplasm. Table 8.3 illustrates this, concentration being expressed relatively in terms of dilutions of a crude culture filtrate. The absolute values are not important for the present purpose, which is to show a quantitative effect; and variations in the potency of crude filtrates can be ignored. It is to be understood that there is no sharp transition as concentration is raised, and toxicity as judged by, for example, the effect on mitochondria being greater than toxicity as judged by external visual symptoms.

Table 8.3. The different reactions of maize with normal and with Texas male sterile (Tms) cytoplasm to varying concentrations of *Helminthosporium maydis* race T toxin[a]

Toxin concentration[b]	Plant cytoplasm	
	Tms	Normal
Less than 1/5000	Resistant	Resistant
Between 1/5000 and 1/200	Susceptible	Resistant
Greater than 1/200	Susceptible	Susceptible

[a] Data of Wheeler (1975).
[b] Concentration in terms of crude culture filtrates, 1/5000 being filtrate diluted with 5000 parts of water, and so on.

Table 8.3 does what Table 8.2 cannot do. It distinguishes the quantitative from the qualitative, and shows that the effect is quantitative. Table 8.3 reflects Case B of Table 8.2. Changes are gradual in the region of the thresholds. Increasing concentrations of toxin increase damage in both sorts of cytoplasm, i.e., the ranking order is the same down the table. Plants with Tms cytoplasm are more damaged than those with normal cytoplasm, i.e., the ranking order is the same across the table. Table 8.3 meets the ranking test for horizontal resistance (Sect. 1.11); in so far as one may substitute the toxin for the pathogen itself, the evidence indicates that the resistance of normal maize cytoplasm is horizontal.

The same test can be applied mutatis mutandis to the other diseases caused by selective pathotoxins and listed in Table 8.1, indicating that the resistance to toxin in all of them is horizontal.

The horizontal resistance is specific and, in most instances, monogenic. The previous chapter prepared the way to accepting this, by disposing of the myths, embodied in badly chosen names, that horizontal resistance is general resistance, or that it is polygenic resistance.

8.5 Horizontal Resistance with a Threshold

For the particular diseases listed in Table 8.1, the horizontal resistance to toxin and, by implication, to disease evidently differs from better known examples of horizontal resistance. The difference is that a threshold is involved. Toxin concentration must reach a threshold value before disease occurs. If it fails to reach this threshold value in the pathogen, the host is resistant. For example, the toxin victorin has never reached the threshold value needed for *Helminthosporium victoriae* to attack Bond or other resistant varieties of oats.

Instead of thinking of threshold levels of toxin in the pathogen, one could equally well think in terms of doses of toxin tolerated by the host. There is nothing new to toxicology in the concept of a tolerated dose of toxin.

Resistance to the diseases listed in Table 8.1 has never been lost. When the Victoria oat was introduced, resistance was not lost. What happened was that a susceptible variety replaced resistant varieties in agriculture, a reverse of the usual procedure of replacing susceptible by resistant varieties. Bond and other resistant varieties have remained as resistant as they ever were. So too the great epidemic of southern maize leaf blight in 1970 occurred because maize with susceptible Tms cytoplasm replaced maize with normal cytoplasm. The resistance of normal cytoplasm is as good as it ever was. In the more usual agricultural procedure of replacing a susceptible variety by a resistant variety, the resistance of sorghum to *Periconia circinata* is as good as it was when it was first introduced.

It is however no part of the concept of horizontal resistance with a threshold to assume that in no host variety will the tolerated dose of toxin ever be exceeded; but it is part of the concept to believe that even if the tolerated dose were exceeded, a resistant variety would still suffer less than a susceptible variety, other genetic factors being equal. Until such time as appropriate examples are found there is no purpose in pursuing this matter further.

Chapter 9 A Molecular Hypothesis of Vertical and Horizontal Resistance

9.1 The Hypothesis

A primary coded protein which is an enzyme or subunit of an enzyme can react in two different ways. It can polymerize (given the necessary hydrophobicity); it can catalyze (given the necessary substrate); or it can do both (when the catalyst is itself a polymer).

Vertical resistance or susceptibility is determined by polymerization. Horizontal resistance or susceptibility is determined by catalysis or the products of catalysis. The hypothesis is as simple as that and it fits the facts.

The one half of the hypothesis, about protein polymerization, follows from what was discussed at length in Chapters 2, 3, and 4. The other half, about catalysis, is explained in the following section. Thereafter, in Sections 9.3, 9.4, 9.5, and 9.6, we are concerned primarily to illustrate the hypothesis and fill in some details.

9.2 Catalysis the Transformer and Eroder of Qualitative Variation

Consider the primary protein coded by DNA. On an average 17/24 of the mutations represented as base substitutions in the DNA are passed on as amino acid substitutions in the protein (see Ch. 2). Because the protein molecule commonly contains several hundred amino acid residues, the scope for variation in the protein is large.

This variation is qualitative. A qualitative change of a DNA base reaches the protein as a qualitative change in its amino acids.

Here a dichotomy enters. If the protein polymerizes, all the qualitative variation is conserved as qualitative variation, and the polymer contains the same variety of amino acids as does its constituent monomers. If the protein catalyzes, almost all the qualitative variation is destroyed, and what passes on to the catalyzed product is quantitative variation, resulting from the different catalytic efficiencies of the enzyme variants. Catalysis is the great transformer of qualitative into quantitative variation. Very often, the amount of quantitative variation passed on to the catalyzed product is small; it may even be absent. This happens when an amino acid substitution in the enzyme does not affect the active site of the enzyme or its general conformation; and the result is that variation greatly decreases or even disappears during catalysis.

Catalysis is not only a great transformer of qualitative into quantitative variation; it is also a great eroder of variation, and the erosion of variation enters modern theories of evolution in the form of discussions about the significance of neutral mutations. Neutral mutations have become manifest in comparisons of amino acid sequences in homologous enzymes of different organisms. It has been found that there is not a single best sequence, but a wide variety of sequences is possible for any particular catalytic function. Many of the amino acid differences apparently confer little or no selective advantage. Within populations of the same organism neutral or near-neutral mutation is also manifest in the abundance of isozymes, which are enzyme variants with a common catalytic function.

Langridge (1974) approached the matter experimentally. He treated *Escherichia coli* with a mutagen (N-methyl-N-nitro-N-nitrosoguanidine) and examined the variation in β-galactosidase activity. Of 733 amino acid mutations calculated to have occurred, only 11 reduced β-galactosidase activity by more than half. The inertness of the enzyme to amino acid replacement was confirmed by immunological tests of β-galactosidase molecules changed in amino acid sequence by suppression.

Catalysis transforms most qualitative variation into quantitative variation, and erodes much of that variation. That is the rule. The exception is qualitative variation that survives catalysis, and remains as qualitative variation. It is relatively very rare.

Nature is conservative about qualitative changes in the products of enzymes. To cite an example of direct relevance to plant pathogenesis, the primary cell walls of a variety of dicotyledons are similar in that they are composed of the same monosaccharide constituents joined together by the same glycosidic linkages (Wilder and Albersheim, 1973; Albersheim and Anderson-Prouty, 1975). So too the monocotyledons, though they differ in some ways from dicotyledons, have among themselves qualitatively similar primary cell walls (Burke et al., 1974; Albersheim and Anderson-Prouty, 1975). There is presumably a great abundance of isozyme variations of the enzymes that lay down these cell walls, but any variations passed on to their products are almost always just quantitative. Within each sub-group of flowering plants, qualitative variation in the primary coded proteins does not reach the primary cell walls, and the walls are left practically invariant qualitatively.

There is good reason for nature's conservatism that freezes qualitative variations. When once a metabolic system has been established that functions well, any qualitative change in an enzyme's products would need simultaneous changes in functionally coupled processes. Mutations being rare and random, the probability that two or more independent gene loci would simultaneously sustain mutually acceptable mutation is vanishingly small; see also Ohno (1973). Quantitative variations, on the other hand, are better tolerated; and enzymes normally have spare catalytic capacity which can compensate for minor changes in efficiency, at least in the short run.

Against this background we can return to our hypothesis.

9.3 Vertical Resistance: Qualitative Variation

If (to repeat a story) one allows for 20 *Sr* genes for resistance to wheat stem rust, one must allow for a potential million phenotypes of wheat varying in their reaction to the pathogen, and a potential million phenotypes of the pathogen varying in their reaction to wheat. The variation in both host and pathogen is qualitative; one can place both host and pathogen in different and discontinuous classes, called races when one thinks of the pathogen, and differential varieties or some such name when one thinks of the host.

This vast mass of qualitative variation (the hypothesis says) does not survive catalysis; it goes only so far as the primary coded protein. This holds for both host and pathogen.

For a differential interaction to be expressed in host–pathogen interactions, e.g., as differential races of the pathogen, the qualitative variation must be conserved in *both* host and pathogen. Therefore (the hypothesis says) the primary coded protein of the host and that of the pathogen must interact directly without any product of catalysis being an intermediary; and that means copolymerization.

9.4 Horizontal Resistance: Quantitative Variation

The results of catalytic reaction are quantitative, even though one of the partners in the reaction varies qualitatively, provided that the other partner varies quantitatively. As an illustration consider the degradation of host cell walls by pathogen enzymes. The quantity of cell wall material affects the ability of the pathogen to degrade the wall (Tomiyama, 1963; Griffey and Leach, 1965; Mercer et al., 1974; Albersheim and Anderson-Prouty, 1975). Thick walls are harder or take more time to degrade, which is possibly why in many diseases relatively high resistance is shown by the older parts of plants. However, the cellulases, polygalacturonases and other enzymes that degrade these walls vary qualitatively as isozymes. The host cellulose, pectins and other wall constituents are likely to have lost, during their catalytic synthesis, most or all of the qualitative variation of their parent enzymes (see Sect. 9.2); and the reaction between quantitatively varying cellulose and qualitatively varying cellulase, quantitatively varying pectins and qualitatively varying pectic enzymes, and so on, must be essentially quantitative. Albersheim and Anderson-Prouty (1975), in their review of the biochemistry of pathogenesis, quote the reaction between cell walls and wall-degrading enzymes as an example of horizontal resistance, as our hypothesis requires.

According to the hypothesis, if a product of catalysis is not involved in horizontal resistance, catalysis itself must be involved. The catalytic digestion of enzymes by enzymes, presumably a common phenomenon in necrotrophic disease, might well provide examples of this.

9.5 Vertical Resistance: Reciprocal Host–Pathogen Variation in the Same Class

From the discussions in Chapters 2 and 3 it was concluded that mutation to resistance in the host involved a change toward a less hydrophobic amino acid side chain, and therefore less polymerization; and that mutation to virulence in the pathogen involved a change towards a more hydrophobic amino acid side chain, and therefore more polymerization. Observe the balance. An amino acid change in the host can be countered by an amino acid change in the pathogen. The changes are of the same class and reciprocal. What the host can do, the pathogen can likewise do. If a plant breeder incorporates resistance in the host, the pathogen can reply in the same kind; hence stems the frailty of vertical resistance; hence stems the correlated variation of pathogen and host that is the basis of vertical resistance.

9.6 Horizontal Resistance: Host–Pathogen Variation of Different Kinds

In horizontal resistance the pathogen cannot reply in kind to a host's change to resistance; hence stems much of the stability of horizontal resistance.

For illustration consider sugarcane eyespot disease, caused by *Helminthosporium sacchari* and discussed in the previous chapter. On the evidence given there, a glycoside from the pathogen binds with a protein in the host. Glycosides and proteins differ entirely in kind and, what is possibly also important, differ in the number of links in the chain of synthesis that starts from the DNA. In more detail, an α-galactoside from the pathogen binds with a protein in susceptible sugarcane clones, presumably, if one may borrow ideas from current lectin protein theory, by fitting into a cleft or slot in the protein. The α-galactoside does not bind with the protein in resistant clones, presumably because with resistance there is not a fitting cleft or slot. If the sugarcane breeder introduces a resistant clone, what can the pathogen do about it? Must it change from a galactoside to a glucoside or a mannoside, or from the α to the β anomeric form? Even if this change were genetically feasible, how would it help the pathogen, if the cleft or slot is gone?

Similar comments probably hold for selective pathotoxin diseases in general, in that it is unlikely that the host's receptor is of the same molecular class as the toxin itself.

When the pathogen's enzymes degrade host cell walls, proteins are matched with polysaccharides, again requiring the pathogen to meet variation in the host with variation of a different molecular kind.

Superficially, it might seem that pathogen and host variation are of the same kind when enzyme degrades enzyme catalytically, but this is not so. Different parts of the molecule are involved in the substrate and the catalyst; and substrate and catalyst are not interchangeable. Variation is not reciprocal.

Chapter 10 Biotrophy, Necrotrophy, and the Lineage of Symbiosis

10.1 Introduction

Table 10.1 shows how diverse are the processes involved in plant pathogenesis. With a cereal rust disease, a plant is resistant if the host cell nucleus dies after inoculation, but susceptible if it lives, at least for a while. With a soft rot disease, a plant is susceptible if the host cell nucleus dies after inoculation, but resistant if it lives.

The zoological sciences have an advantage over us. They distinguish between parasitism and predation. There are still parts of the world where you can die of malaria fever or be eaten by a lion. By having two coupled word pairs, host/parasite and prey/predator, the reality of the difference is better brought out. In plant pathology, with only a single word pair, host/parasite (or host/pathogen) we perforce smother essential differences.

A cereal rust disease is commonly called a biotrophic disease, and a soft rot a necrotrophic disease. The terms, parasitic symbiosis and asymbiosis, might be better. Table 10.1 refers simply to the host cell nucleus. This puts the nucleus at the center of events, as is proper. At the same time it allows for all gradations, from an indefinitely preserved host nucleus, as in many virus diseases, to a nucleus killed in advance of the parasite, as in many soft rot diseases.

Against this background we return to the topic of the previous chapter. All diseases involve enzymes as catalysts. Against all disease, be it a cereal rust or a soft rot, we can expect horizontal resistance, in some degree or other. But protein polymerization is likely to occur extensively only with a continuous supply of protein. The nucleus determines this supply, and vertical resistance based on protein polymerization is likely to be important only in biotrophic associations. The facts fit, and there is as yet no evidence for vertical resistance against soft rots.

In the cereal rusts and other diseases with highly developed parasitic symbiosis the parasite makes the host deliver food. Proteins, it seems, are delivered

Table 10.1. Resistance and susceptibility of the host plant in relation to the behavior of the host cell nucleus after inoculation

Disease	Nucleus lives[a]	Nucleus dies
Cereal rust	Susceptible	Resistant
Soft rot	Resistant	Susceptible

[a] Nucleus lives at least for a while.

mainly as such, rather than broken down as amino acids; in pathogenesis the abundance of RNA, ribosomes, and other features of protein synthesis point that way. Accept this, and one must inquire how in parasitic symbiosis the parasite gets its protein without destroying the cytoplasm and nucleus which are its source. Proteolytic enzymes as digestive ferments are indiscriminately destructive, and are incompatible with a living nucleus and cytoplasm. We look elsewhere, and find the answer in protein copolymerization.

There is nothing new in the concept that in necrotrophic parasitism the pathogen releases enzymes that destroy and digest the host cell, whereas in biotrophy enzyme release must be controlled so as to avert cell destruction. The concept pervades the literature and is practically axiomatic.

De Bary (1887) divided fungi into three groups based on how they fed: saprophytes never parasitic, facultative parasites usually saprophytic but sometimes parasitic, and obligate parasites needing parasitism for their full development. This division, together with various subdivisions, has resulted in speculation about the origin of biotrophy in a chain of nutritional sequences that starts with saprophytism or necrotrophy and ends in obligate parasitism. Lewis (1973), in a review dealing essentially with carbohydrates, has suggested such a chain, with biotrophy arising from necrotrophy by the suppression of degradative enzymes and the concurrent production of hormones that permit infected tissues to import products of photosynthesis. Cain (1972), in an essay on the evolution of the fungi, also provided for a gradual and orderly evolution starting with saprophytism, and extended his thesis to suggest that, in order to explain some of the peculiar host–pathogen relationships, it may be necessary to conclude that the parasitic habit in rust fungi developed from the saprophytic habit independently in more than one species.

Theories that biotrophy originated in necrotrophy or saprophytism leave too much to be explained. In *Triticum* there are probably at least 50 genes matched gene-for-gene by those of three *Puccinia* spp. in an essentially biotropic association. Where did these genes of *Puccinia* come from? There is nothing to suggest a necrotrophic background. Indeed, the genes are concerned in the nutrition of the parasite only when necrosis is substantially absent. That is a clear message from Table 10.1.

10.2 Necrotrophy, Biotrophy, and Biotrophy-Necrotrophy Sequences

We see biotrophy and necrotrophy differently. They are separate phenomena, derived separately from separate origins. Biotrophy does not come from necrotrophy, nor necrotrophy from biotrophy. Biotrophic parasitism could exist, even if necrotrophic parasitism had never existed; and necrotrophic parasitism could exist, even if biotrophic parasitism had never existed. Further, biotrophy and necrotrophy are not mutually exclusive; and necrotrophy can replace biotrophy in sequence during lesion development.

The argument goes back to the nature of susceptibility to disease. The misnamed genes for susceptibility are not genes of invitation to the parasite, but are normal plant genes with a useful and necessary function within the healthy plant. This was discussed in Section 2.4 in relation to gene-for-gene diseases, on arguments based largely on principles of population genetics. The same arguments apply, however, to other diseases as well (with the possible and rather doubtful exception of diseases caused by predators like *Sclerotium rolfsii* that attack from a food base). We see plants as having genes, normal to their healthy state, which the parasite uses for its own purposes. Biotrophy and necrotrophy being biochemically distinct, the host genes relevant to them are almost certainly also distinct. Susceptibility genes permitting biotrophy permit symbiosis; those permitting necrotrophy do not. The inference therefore is that there are at least two sorts of susceptibility genes, both normal to the healthy plant but exploited differently by the parasite.

When necrotrophy replaces biotrophy within a lesion, the genes controlling the enzymes and toxins associated with necrotrophy are presumably repressed during the phase of biotrophy, and then derepressed when necrotrophy takes over. Such a notion, that lesion development requires the repression and derepression of genes at appropriate times in appropriate sequence, is compatible with the concepts of developmental genetics. To give an example, the invasion of leaves of susceptible potato varieties by *Phytophthora infestans* is at first biotrophic and parasitically symbiotic. The nuclei of affected host cells swell but remain alive and functionally active for several days, until sporulation by the fungus is well established. Thereafter the process becomes necrotrophic; sporulation stops, and the older parts of the lesion turn necrotic. Presumably in *P. infestans* (though not in the more highly biotrophic cereal rust fungi) the change from biotrophy to necrotrophy occurs because the fungus, by derepressing the appropriate genes, starts to release digestive and therefore destructive enzymes to meet the nutritional demands of sporulation. In terms of our hypothesis, the genes involved in the biotrophic phase and those in the necrotrophic phase are different, and are manipulated in appropriate sequence to suit the pathogen's needs.

Initial biotrophy, even if it is only transient, is of great practical importance because it allows vertical resistance to occur even when disease in a lesion ends necrotrophically.

Pure necrotrophy outside of the soft rots may perhaps be rare in plant disease. Even diseases associated with selective pathotoxins may have a biotrophy-necrotrophy sequence; Wheeler (1977) has shown that the pathotoxins of *Helminthosporium maydis* race T play no detectable role in pathogenesis during an interval immediately after infection is established. Because biotrophy may make way for necrotrophy in a lesion, the classification of pathogens as biotrophic or necrotrophic must be done with caution, even when the classification of *processes* as biotrophic or necrotrophic is permissible. In classifying pathogens, one is on safer ground if one relies on factual observation and bases the classification on the survival of the cell nucleus of susceptible (compatible) host plants. This system, used broadly in Table 10.1, recognizes all gradations that sequential change from necrotrophy to biotrophy may bring.

10.3 The Lineage of Symbiosis

Biotrophy may well have its origin in symbiosis. Put more pertinently, parasitic symbiosis and mutualistic symbiosis may well be associated, and this could give biotrophy a very ancient lineage.

Hallbauer and van Warmelo (1974) and Hallbauer et al. (1977) have described a Precambrian plant 2500 million years old from the carbonaceous matter in the Witwatersrand gold mines. It is *Thuchomyces lichenoides* Hallbauer and Johns, illustrated in the frontispiece. (The holotype is in the Senckenberg Museum, Frankfurt, Germany, Nr. S.M.B. 12794/1,2.) The anatomical structure as preserved by encrustation with gold is remarkably like that of living lichens such as *Parmelia*. Here is a hint of symbiosis existing far back in time. Evidence from various sources suggests that the atmosphere then was very low in oxygen; and Echlin (1966) and Hallbauer and van Warmelo (1974) have pointed out the advantages in anaerobic conditions of an association between a photosynthetic alga and a nonphotosynthetic filamentous organism, as in a lichen. We can keep an open mind about details of events and organisms of 2500 million years ago, but still accept evidence that symbiosis in plants may be very old. Echlin (1966), too, postulates symbiosis in Precambrian times.

The hypothesis of protein nutrition through copolymerization may need to be supported by evidence of parallel evolution of host and parasite. Protein copolymerization is highly selective, and the monomers concerned are usually similar in function and structure. The proteins from flagella, the flagellins, illustrate this. Kuroda (1972) found that the flagellins of *Salmonella* and *Proteus* copolymerized in vitro, but those of *Salmonella* and *Bacillus* did not, nor did those of *Proteus* and *Bacillus*. The requirements for copolymerization are evidently very precise.

It may be necessary to postulate that the relevant protein metabolisms of host and parasite were originally one, and that they have diverged through gene duplication followed by mutation. The theory of evolution by gene duplication has been discussed by Ohno (1970). (It will be remembered from Sect. 6.8 that gene duplication followed by mutation was used to explain nonallelic interaction between virulences.) All that needs to be added to the theory in order to adapt it to symbiosis is the realistic suggestion that selection pressure kept the relevant mutations in the pathogen aligned with those in the host. One of the interesting possibilities not so far-fetched as might appear at first thoughts is that the fungi were the original eukaryotes, the ancestors of us all. (Of all organisms the fungi make the most bridges between eukaryotic animals and plants. As has often been remarked in the literature, they use the same sort of food as animals; they make chitin; some have amoeboid movement; some are unicellular; and so on. On the theory widely discussed in botanical literature that chloroplasts were originally endosymbiotic prokaryotic algae, green plants might well have started from genetic and molecular associations like those still surviving in lichens.) If this possibility is correct, *Triticum* and *Puccinia* may share more ancestral genes than is superficially apparent.

The two great classes of symbiotic fungus/higher plant associations are those of parasites substantially biotrophic and of endogenous mycorrhizae. Both classes occur ubiquitously, and may be more closely related than superficial differences suggest. The most striking difference outside of ecological adaptation is in host–parasite specificity. The rusts, powdery mildews and most other mainly biotrophic parasites exist as a multitude of races each with a narrow host range; paraphrased, there are a multitude of resistance barriers thrown up against them. The vesicular-arbuscular mycorrhizal fungi (*Endogone* spp.) have wide host ranges and few races; paraphrased, there are few resistance barriers against them. The difference in host–parasite specificity is easily explained against two backgrounds. First, it is the allele for resistance that determines specificity, but the allele for susceptibility that permits symbiosis. Second, in relation to resistance, selection pressures in parasitic symbiosis and those in mutualistic symbiosis are in precisely opposite directions. Consider these matters in more detail.

We recognize relevant gene loci through the resistance allele. Flor's gene-for-gene hypothesis, which is a hypothesis about host–parasite specificity, could not have existed but for resistance alleles, and concerns only those loci which have resistance alleles. We identify and count the *Sr* alleles in wheat for resistance to stem rust; and there may be many more *sr* alleles for susceptibility about which we know nothing (not even whether they exist) because there has been no mutation to resistance. On the other hand, susceptibility, which in biotrophy is symbiosis, is determined by the absence of alleles for resistance. In parasitic symbiosis it is the *sr* not the *Sr*, the *lr* not the *Lr* alleles that are involved. So too mutualistic symbiosis, as with the mycorrhizae, depends on genes for "susceptibility", and the rarity of fungus races in mutualistic symbiosis reflects the rarity of genes for "resistance" in the "host" plant.

Opposite selection pressures are clearly involved. In parasitic symbiosis the host plant benefits by mutation to resistance because this ends an (for the host) unwanted symbiosis. In mutualistic symbiosis the "host" plant loses by mutation to resistance, because this ends the wanted symbiosis. Mutations to resistance in mycorrhizal host plants are eliminated by selection because they are disadvantageous; and this elimination also eliminates a major source of specificity.

References

Abbott, L. K.: Taxonomy and host specificity of Ophiobolus graminis Sacc.; an application of electrophoretic and serological techniques. Ph. D. thesis, Monash Univ., Victoria, Australia (1973). Quoted in De Vay, J. E.: Protein specificity in plant disease development: protein sharing between host and parasite. In: Wood, R. K. S., Graniti, A. (Eds.): Specificity in Plant Diseases, pp. 199–212. New York: Plenum 1975

Albersheim, P., Anderson-Prouty, A. J.: Carbohydrates, proteins, cell surfaces, and the biochemistry of pathogenesis. Ann. Rev. Plant Physiol. **26**, 31–52 (1975)

Allen, R. F.: A cytological study of Puccinia recondita physiological form 11 on Little Club wheat. J. Agric. Res. **33**, 201–222 (1926)

Anderson, R. G.: The inheritance of leaf rust resistance in seven varieties of common wheat. Can. J. Plant Sci. **41**, 342–359 (1961)

Antonelli, E., Daly, J. M.: Decarboxylation of indoleacetic acid by near isogenic lines of wheat resistant or susceptible to Puccinia graminis f. sp. tritici. Phytopathology **56**, 610–618 (1966)

Ayers, A. R., Ebel, J., Finelli, F., Berger, N., Albersheim, P.: Host-pathogen interactions. IX. Quantitative assays of elicitor activity and characterization of the elicitor present in the extracellular medium of cultures of Phytophthora megasperma var. sojae. Plant Physiol. **57**, 751–759 (1976a)

Ayers, A. R., Ebel, J., Valent, B., Albersheim, P.: Host-pathogen interactions. X. Fractionation and biological activity of an elicitor isolated from the mycelial walls of Phytophthora megasperma var. sojae. Plant Physiol. **57**, 760–765 (1976b)

Ayers, A. R., Valent, B., Ebel, J., Albersheim, P.: Host-pathogen interactions. XI. Composition and structure of wall-released elicitor fractions. Plant Physiol. **57**, 766–774 (1976c)

Bartos, P., Dyck, P. L., Samborski, D. J.: Adult-plant leaf rust resistance in Thatcher and Marquis wheat: a genetic analysis of the host-parasite situation. Can. J. Bot. **47**, 267–269 (1969)

Bary, A., de: Comparative Morphology and Biology of the Fungi, Mycetozoa and Bacteria. Oxford: Clarendon 1887

Biffen, R. H.: Mendel's laws of inheritance and wheat breeding. J. Agric. Sci. **1**, 4–48 (1905)

Black, W., Mastenbroek, C., Mills, W. R., Peterson, L. C.: A proposal for an international nomenclature of races of Phytophthora infestans and of genes controlling immunity in Solanum demissum derivatives. Euphytica **2**, 173–179 (1953)

Bohlool, B. B., Schmidt, E. L.: Lectins: a possible basis for specificity in Rhizobium-legume root-nodule symbiosis. Science **185**, 269–271 (1974)

Bonde, R., Schultz, E. S.: Potato refuse piles as a factor in the dissemination of late blight. Maine Agric. Exp. Stn. Bull. **416**, 230–246 (1943)

Bonde, R., Schultz, E. S.: Potato refuse piles and late-blight epidemics. Maine Agric. Exp. Stn. Bull. **426**, 233–234 (1944)

Brinkerhoff, L. A.: Variation in Xanthomonas malvacearum and its relation to control. Ann. Rev. Phytopathol. **8**, 85–110 (1970)

Brinkerhoff, L. A., Presley, J. T.: Effect of four day and night temperatures regimes on bacterial blight of immune, resistant and susceptible strains of upland cotton. Phytopathology **57**, 47–51 (1967)

Bromfield, K. R.: The effect of postinoculation temperature on seedling reaction of selected wheat varieties to stem rust. Phytopathology **51**, 590–593 (1961)

Bugbee, W. M., Sappenfield, W. P.: Varietal reaction in cotton after stem and root inoculation with Fusarium oxysporum f. sp. vasinfectum. Phytopathology **58**, 212–214 (1968)

Burke, D., Kaufman, P., McNeil, M., Albersheim, P.: The structure of plant cell walls. VI. A survey of walls of suspension-cultured monocots. Plant Physiol. **54**, 109–115 (1974)

Burrows, V. D.: Nature of resistance of oat varieties to oat stem rust. Nature (London) **188**, 957–958 (1960)

Butler, E. J., Jones, S. G.: Plant Pathology. London: Macmillan 1949

Cain, R. F.: Evolution of the fungi. Mycologia **64**, 1–14 (1972)

Calhoun, D. H., Hatfield, G. W.: Autoregulation of gene expression. Ann. Rev. Microbiol. **29**, 275–299 (1975)

Cammack, R. H.: Puccinia polysora: a review of some factors affecting the epiphytotic in West Africa. Rept. 6th Commonwealth Mycological Conf. 1960, 134–138 (1961)

Carleton, M. A.: Cereal rusts of the United States: a physiological investigation. U.S. Dep. Agric. Div. Vegetable Physiol. Pathol. Bull. **16** (1899)

Cason, E. T., Richardson, P. E., Brinkerhoff, L. A., Gholson, R. K.: Histopathology of immune and susceptible cotton cultivars inoculated with Xanthomonas malvacearum. Phytopathology **67**, 195–198 (1977)

Caten, C. E.: Intra-racial variation in Phytophthora infestans and adaptation to field resistance for potato blight. Ann. Appl. Biol. **77**, 259–270 (1974)

Charudattan, R., De Vay, J. E.: Common antigens among varieties of Gossypium hirsutum and isolates of Fusarium and Verticillium species. Phytopathology **62**, 230–234 (1972)

Chester, K. S.: The problem of acquired physiological immunity in plants. Q. Rev. Biol. **8**, 275–324 (1933)

Chester, K. S.: The Nature and Prevention of the Cereal Rusts as Exemplified in the Leaf Rust of Wheat. Waltham, Massachusetts: Chronica Botanica 1946

Chothia, C., Janin, J.: Principles of protein-protein recognition. Nature (London) **256**, 705–708 (1975)

Cirulli, M., Ciccarese, F.: Interactions between TMV isolates, temperature, allelic condition and combination of the *Tm* resistance genes in tomato. Phytopathol. Mediterr. **14**, 100–105 (1975)

Clarke, A. E., Harrison, S., Knox, R. B., Raff, J., Smith, P., Marchalonis, J. J.: Common antigens in male-female recognition in plants. Nature (London) **265**, 161–162 (1977)

Clifford, B. C.: The histology of race non-specific resistance to Puccinia hordei Otth. in barley. Proc. 5th European and Mediterranean Cereal Rust Conf., Prague, **1**, 75–79 (1972)

Clifford, B. C., Clothier, R. B.: Physiologic specialization of Puccinia hordei on barley hosts with non-hypersensitive resistance. Trans. Br. Mycol. Soc. **63**, 421–430 (1974)

Coffey, M. D.: Flax rust resistance involving the K gene: an ultrastructural survey. Can. J. Bot. **54**, 1443–1457 (1976)

Cross, J. E.: Pathogenicity differences in Tanganyika populations of Xanthomonas malvacearum. Empire Cotton Growing Rev. **40**, 125–130 (1963)

Cross, J. E.: Field differences in pathogenicity between Tanganyika populations of Xanthomonas malvacearum. Empire Cotton Growing Rev. **41**, 44–48 (1964)

Crosse, J. E.: Variation amongst plant pathogenic bacteria. Ann. Appl. Biol. **81**, 438 (1975)

Daly, J. M.: The influence of nitrogen source on the development of stem rust of wheat. Phytopathology **39**, 386–393 (1949)

Daly, J. M.: The use of near-isogenic lines in biochemical studies of the resistance of wheat to stem rust. Phytopathology **62**, 392–400 (1972)

Daly, J. M., Ludden, P., Seevers, P.: Biochemical comparisons of resistance to wheat stem rust disease controlled by the Sr 6 and Sr 11 alleles. Physiol. Plant Pathol. **1**, 397–407 (1971)

Damian, R. T.: Molecular mimicry: Antigen sharing by parasite and host and its consequences. Am. Nat. **98**, 129–149 (1964)

Danielli, J. F., Davson, H.: A contribution to the theory of permeability of thin films. J. Cell Physiol. **5**, 495–508 (1935)

Davidson, W. D.: The rejuvenation of the Champion potato. Econ. Proc. R. Dublin Soc. **21**, 319–330 (1928)

Day, P. R.: Genetics of Host-Parasite Interaction. San Francisco: Freeman 1974

De Vay, J. E., Charudattan, R., Wimalajeewa, D. L. S.: Common antigenic determinants as a possible regulator of host-pathogen compatibility. Am. Nat. **106**, 185–194 (1972)

De Vay, J. E., Schnathorst, W. C., Foda, M. S.: Common antigens and host-parasite interactions. In: Mirocha, C., Uritani, L. (Eds.): The Dynamic Role of Molecular Constituents in Plant-parasitic Interactions, pp. 313–328. St. Paul: Bruce Publishing 1967

Dineen, J. K.: Antigenic relationship between host and parasite. Nature (London) **197**, 471–472 (1963)

Doke, N., Tomiyama, K.: Effect of blasticidin S on hypersensitive death of potato leaf petiole cells caused by infection with an incompatible race of Phytophthora infestans. Physiol. Plant Pathol. **6**, 169–175 (1975)

Doubly, J. A., Flor, H. H., Clagett, C. O.: Relation of antigens of Melampsora lini and Linum usitatissimum to resistance and susceptibility. Science **131**, 229 (1960)

Dropkin, V. H.: A necrotic reaction of tomatoes and other hosts resistant to Meloidogyne: Reversal by temperature. Phytopathology **59**, 1632–1637 (1969)

Durrell, L. W., Parker, J. H.: Comparative resistance of varieties of oats to crown and stem rusts. Iowa Agric. Exp. Stn. Res. Bull. **62** (1920)

Dyck, P. L., Samborski, D. J., Anderson, R. G.: Inheritance of adult-plant leaf rust resistance derived from the common wheat varieties Exchange and Frontana. Can. J. Genet. Cytol. **8**, 665–671 (1966)

Ebel, J., Ayers, A. R., Albersheim, P.: Host-pathogen interactions. XII. Response of suspension-cultured soybean cells to the elicitor isolated from Phytophthora megasperma var. sojae, a fungal pathogen of soybeans. Plant Physiol. **57**, 775–779 (1976)

Echlin, P.: Origins of photosynthesis. Science J. **2**, 42–47 (1966)

Elliott, C.: Manual of Bacterial Plant Pathogens. Waltham, Mass.: Chronica Botanica 1951

Fedotova, T. I.: Significance of individual proteins of seed in the manifestation of the resistance of plants to diseases. Trudy Leningr. Inst. Zasch. Rast. Sb. **1**, 61–71 (1948)

Fisher, R. A.: The Genetical Theory of Natural Selection. Oxford: Clarendon 1930

Fitch, W. M.: An improved method of testing for evolutionary homology. J. Mol. Biol. **16**, 9–16 (1966)

Flor, H. H.: Inheritance of pathogenicity in Melampsora lini. Phytopathology **32**, 653–669 (1942)

Flor, H. H.: The complementary genic systems in flax rust. Adv. Genet. **8**, 29–54 (1956)

Flor, H. H.: Genetic controls of host-parasite interactions in rust diseases. In: Holton, C. S., Fischer, G. W., Fulton, R. W., Hart, H., McCallan, S. E. A. (Eds.): Plant Pathology: Problems and Progress 1908–1958, pp. 132–144. Madison: Univ. of Wisconsin Press 1959

Flor, H. H.: Current status of the gene-for-gene concept. Ann. Rev. Phytopathol. **9**, 275–296 (1971)

Futrell, M. C., Dickson, J. G.: The influence of temperature on the development of powdery mildew of spring wheats. Phytopathology **44**, 247–251 (1954)

Garber, R. H., Houston, B. R.: Penetration and development of Verticillium albo-atrum in the cotton plant. Phytopathology **56**, 1121–1126 (1966)

Garber, R. H., Houston, B. R.: Nature of Verticillium wilt resistance in cotton. Phytopathology **57**, 885–888 (1967)

Gassner, G., Straib, W.: Zur Frage der Konstanz des Infektionstypus von Puccinia triticina Erikss. Phytopathol. Z. **4**, 57–64 (1932)

Gassner, G., Straib, W.: Experimentelle Untersuchungen zur Epidemiologie des Gelbrostes (Puccinia glumarum [Schm.] Erikss. und Henn.). Phytopathol. Z. **7**, 285–307 (1934)

Goldberger, R. F.: Autogenous regulation of gene expression. Science **183**, 810–816 (1974)

Golik, I. V., Gromova, B. B.-O., Fedotova, T. I.: Immunochemical similarities of proteins of Synchytrium endobioticum (Schilb) Perc. and the host plant. Abstr. Rev. Plant Pathol. **56**, 84–85 (1977)

Gordon, W. L.: Effect of temperature on host reactions to physiologic forms of Puccinia graminis avenae Erikss. and Henn. Sci. Agric. **11**, 95–103 (1930)

Gordon, W. L.: A study of the relation of environment to the development of the uredinal and telial stages of the physiologic forms of Puccinia graminis avenae Erikss. and Henn. Sci. Agric. **14**, 184–237 (1933)

Gough, F. J., Merkle, O. G.: Inheritance of stem and leaf rust resistance in Agent and Agrus cultivars of Triticum aestivum. Phytopathology **61**, 1501–1505 (1971)

Green, G. J.: A color mutation, its inheritance, and the inheritance of pathogenicity in Puccinia graminis Pers. Can. J. Bot. **42**, 1653–1664 (1964)

Green, G. J.: Selfing studies with races 10 and 11 of wheat stem rust. Can. J. Bot. **44**, 1255–1260 (1966)

Green, G. J.: Stem rust of wheat, barley and rye in Canada in 1970. Can. Plant Dis. Surv. **51**, 20–23 (1971a)

Green, G. J.: Physiological races of wheat stem rust in Canada from 1919 to 1969. Can. J. Bot. **49**, 1575–1588 (1971b)

Green, G. J.: Stem rust of wheat, barley and rye in Canada in 1971. Can. Plant Dis. Surv. **52**, 11–14 (1972a)

Green, G. L.: Stem rust of wheat, barley and rye in Canada in 1972. Can. Plant Dis. Surv. **52**, 162–167 (1972b)

Green, G. J.: Stem rust of wheat, barley and rye in Canada in 1973. Can. Plant Dis. Surv. **54**, 11–16 (1974)

Green, G. J.: Stem rust of wheat, barley and rye in Canada in 1974. Can. Plant Dis. Surv. **55**, 51–57 (1975)

Green, G. J.: Stem rust of wheat, barley and rye in Canada in 1975. Can. Plant Dis. Surv. **56**, 15–18 (1976a)

Green, G. J.: Adult plant reactions of commercial varieties of common wheat to new races of stem rust identified in 1974. Can. Plant Dis. Surv. **56**, 46–47 (1976b)

Griffey, R. T., Leach, J. G.: The influence of age of tissue on the development of bean anthracnose lesions. Phytopathology **55**, 915–918 (1965)

Grümmer, G., Günter, E., Eggert, D.: Die Prüfung von Tomatensorten und ihren Hybriden auf Blatt- und Fruchtbefall mit Phytophthora infestans. Theor. Appl. Genet. **39**, 232–238 (1969)

Hadidi, A., Fraenkel-Conrat, H.: Characterization and specificity of soluble RNA polymerase of brome mosaic virus. Virology **52**, 363–372 (1973)

Hallbauer, D. K., Jahns, H. M., Beltmann, H. A.: Morphological and anatomical observations on some precambrian plants from the Witwatersrand, South Africa. Geol. Rundschau **66**, 477–491 (1977)

Hallbauer, D. K., van Warmelo, K. T.: Fossilized plants in thucholite from precambrian rocks of the Witwatersrand, South Africa. Precambrian Res. **1**, 199–212 (1974)

Hariharasubramanian, V., Hadidi, A., Singer, B., Fraenkel-Conrat, H.: Possible identification of a protein in brome mosaic virus infected barley as a compound of viral RNA replicase. Virology **54**, 190–198 (1973)

Hart, H.: Nature and variability of disease resistance in plants. Ann. Rev. Microbiol. **3**, 289–316 (1949)

Hassebrauk, K.: Untersuchungen über den Einfluß einiger Außenfaktoren auf das Anfalligkeitverhalten der Standardsorten gegenüber verschiedenen physiologischen Rassen des Weizenbraunrostes. Phytopathol. Z. **12**, 233–276 (1940)

Hingorani, M. K.: Factors affecting the survival ability of certain physiological races of Puccinia graminis avenae E. and H. Doctoral thesis, Univ. of Minnesota (1947). Quoted by Hart (1949)

Hislop, E. C., Stahmann, M. A.: Peroxidase and ethylene production by barley leaves infected with Erysiphe graminis f. sp. hordei. Physiol. Plant Pathol. **1**, 297–312 (1971)

Hogen Esch, J. A., Zingstra, H.: Geniteurslijst voor Aartappelrassen. Wageningen: Commissie ter Bevordering van het Kweken en het Onderzoek van nieuwe Aartappelrassen, 1957

Hohl, H. R., Suter, E.: Host-parasite interfaces in a resistant and a susceptible cultivar of Solanum tuberosum inoculated with Phytophthora infestans: leaf tissue. Can. J. Bot. **54**, 1956–1970 (1976)

Holde, K. E. van: The molecular architecture of multichain proteins. In: Molecular Architecture in Cell Physiology, Symp. Soc. Cell Physiol., pp. 89–96. New York: Prentice Hall 1966

Holtzmann, O. V.: Effect of soil temperature on resistance to root-knot nematodes (Meloidogyne incognita). Phytopathology **55**, 990–992 (1965)

Hooker, A. L.: Monogenic resistance in Zea mays L. to Helminthosporium turcicum. Crop Sci. **3**, 381–383 (1963)

Hooker, A. L.: The genetics and expression of resistance in plants to rusts of the genus Puccinia. Ann. Rev. Phytopathol. **5**, 163–182 (1967)

Howard, H. W.: The relation between resistance genes in potatoes and pathotypes of potato-root eelworm (Heterodera rostochiensis), wart disease (Synchytrium endobioticum) and potato virus X. Abst. 1st Int. Congr. Plant Pathol. (London) **92** (1968)

Howes, N. K., Samborski, D. J., Rohringer, R.: Production and bioassay of gene-specific RNA determining resistance of wheat to stem rust. Can. J. Bot. **52**, 2489–2497 (1974)

Hughes, G. R., Hooker, A. L.: Genes conditioning resistance to northern leaf blight in maize. Crop. Sci. **11**, 180–184 (1971)

Huijsman, C. A., Klinkenberg, C. H., Den Ouden, H.: Tolerance to Heterodera rostochiensis Woll. among potato varieties and its relation to certain characteristics of root anatomy. Eur. Potato J. **12**, 134–147 (1969)

Hunter, T. R., Hunt, T., Knowland, J., Zimmern, D.: Messenger RNA for the coat protein of tobacco mosaic virus. Nature (London) **260**, 759–764 (1976)

Ibrahim, I.: Effect of some conditions and chemicals on the development of races 2, 6, 7, and 8 of Puccinia graminis avenae E. and H. Master's thesis, Univ. of Minnesota (1949). Quoted by Hart (1949)

Inoué, S.: Motility of cilia and the mechanism of mitosis. Rev. Mod. Phys. **31**, 402–408 (1959)

Jaenicke, R., Lauffer, M. A.: Polymerization-depolymerization of tobacco mosaic virus protein. XII. Further studies in the role of water. Biochemistry **8**, 3083–3092 (1969)

Jeffrey, S. I. B., Jinks, J. L., Grindle, M.: Interracial variation in Phytophthora infestans and field resistance to potato blight. Genetica **32**, 323–328 (1962)

Jenkins, M. T., Robert, A. L., Findley, W. R.: Recurrent selection as a method for concentrating genes for resistance to Helminthosporium turcicum leaf blight in corn. Agron. J. **46**, 89–94 (1954)

Jinks, J. L., Grindle, M.: Changes introduced by training in Phytophthora infestans. Heredity **18**, 245–264 (1963)

Johnson, L. B., Schafer, J. F.: Identification of wheat leaf rust resistance combinations by differential temperature effects. Plant Dis. Rep. **49**, 222–224 (1965)

Johnson, R., Law, C. N.: Genetic control of durable resistance to yellow rust (Puccinia striiformis) in the wheat cultivar Hybride de Bersée. Ann. Appl. Biol. **81**, 385–391 (1975)

Johnson, R., Taylor, A. J.: Spore yield of pathogens in investigations of race-specificity of host resistance. Ann. Rev. Phytopathol. **14**, 97–119 (1976)

Johnson, T.: Studies in cereal diseases. VI. A study of the effect of environmental factors on the variability of physiologic forms of Puccinia graminis tritici Erikss. and Henn. Can. Dep. Agric. Bull. **140**, NS. (1931)

Johnson, T., Johnson, O.: Studies on the nature of disease resistance in cereals. II. The relationship between sugar content and reaction to stem rust in mature and immature tissues of the wheat plant. Can. J. Res. **11**, 582–588 (1934)

Johnson, T., Newton, M.: The effect of high temperature on uredial development in cereal rusts. Can. J. Res. **15**, 425–432 (1937)

Jones, D. R., Deverall, B. J.: The effect of the *Lr* 20 resistance gene in wheat on the development of leaf rust, Puccinia recondita. Physiol. Plant Pathol. **10**, 275–285 (1977)

Jones, F. G. W., Parrott, D. M.: The genetic relationships of pathotypes of Heterodera rostochiensis Woll. which reproduce on hybrid potatoes with genes for resistance. Ann. Appl. Biol. **56**, 27–36 (1965)

Kamen, R.: Characterization of the subunits of Qβ replicase. Nature (London) **228**, 527–533 (1970)

Kanazawa, Y., Shichi, H., Uritani, I.: Biosynthesis of peroxidase in sliced or black rot-infected sweet potato roots. Agric. Biol. Chem. **29**, 840–847 (1965)

Kaneko, I., Sato, H., Ukita, T.: Effects of metabolic inhibitors on the agglutination of tumour cells by concanavalin A and Ricinus communis agglutinin. Biochem. Biophys. Res. Commun. **50**, 1087–1094 (1973)

Kao, K. N., Knott, D. R.: The inheritance of pathogenicity in races 111 and 29 of wheat stem rust. Can. J. Genet. Cytol. **11**, 266–274 (1969)

Karr, A. L., Karr, D. B., Strobel, G. A.: Isolation and partial characterization of four host-specific toxins of Helminthosporium maydis (race T). Plant Physiol. **53**, 250–257 (1974)

Kassanis, B., Bastow, C.: The relative concentration of infective intact virus and RNA of four strains of tobacco mosaic virus as influenced by temperature. J. Gen. Virol. **11**, 157–170 (1971)

Katsuya, K., Green, G. J.: Reproductive potentials of races 15B and 56 of wheat stem rust. Can. J. Bot. **45**, 1077–1091 (1967)

Kauzmann, W.: Some factors in the interpretation of protein denaturation. Adv. Protein Chem. **14**, 1–63 (1959)

Kawashima, N., Hyodo, H., Uritani, I.: Investigations of antigenic components produced by sweet potato roots in response to black rot infection. Phytopathology **54**, 1086–1092 (1964)

Kawashima, N., Uritani, I.: Occurrence of peroxidase in sweet potato infected by the black rot pathogen. Agric. Biol. Chem. **27**, 409–417 (1963)

Keen, N. T.: Hydroxyphaseollin production by soybeans resistant and susceptible to Phytophthora megasperma var. sojae. Physiol. Plant Pathol. **1**, 265–275 (1971)

Keen, N. T., Partridge, J. E., Zaki, A. I.: Pathogen—produced elicitor of a chemical defense mechanism in soybeans monogenically resistant to Phytophthora megasperma var. sojae. Phytopathology **62**, 768 (1972)

Kim, S. K., Brewbaker, J. L.: Inheritance of general resistance in maize to Puccinia sorghi. Crop Sci. **17**, 456–461 (1977)

Kirschner, K., Bisswanger, H.: Multifunctional proteins. Ann. Rev. Biochem. **45**, 143–166 (1976)

Kirste: Ergebnisse von Krautfäule-Spritzversuchen. Kartoffelbau **9**, 114–115 (1958)

Klement, Z., Farkas, G. L., Lovrekovich, L.: Hypersensitive reaction induced by phytopathogenic bacteria in the tobacco leaf. Phytopathology **54**, 474–477 (1964)

Klotz, I. M.: Comparison of molecular structures of proteins: helix content; distribution of apolar residues. Arch. Biochem. Biophys. **138**, 704–706 (1970)

Knott, D. R., Srivastava, J. P.: Inheritance of resistance to stem rust races 15B and 56 in eight cultivars of common wheat. Can. J. Plant Sci. **57**, 633–641 (1977)

Kondo, M., Gallerani, R., Weissmann, C.: Subunit structure of $Q\beta$ replicase. Nature (London) **228**, 525–527 (1970)

Kosuge, T.: The role of phenolics in response to infection. Ann. Rev. Phytopathol. **7**, 195–222 (1969)

Kuroda, H.: Polymerization of Salmonella, Proteus and Bacillus flagellins in vitro. Biochem. Biophys. Acta **285**, 253–267 (1972)

Lakso, J. U., Starr, M. P.: Comparative injuriousness to plants of Erwinia spp. and other enterobacteria from plants and animals. J. Appl. Bact. **33**, 692–707 (1970)

Langridge, J.: Mutation spectra and neutrality of mutations. Aust. J. Biol. Sci. **27**, 309–319 (1974)

Lapwood, D. H.: Potato haulm resistance to Phytophthora infestans. I. Field assessment of resistance. Ann. Appl. Biol. **49**, 140–151 (1961 a)

Lapwood, D. H.: Potato haulm resistance to Phytophthora infestans. II. Lesion production and sporulation. Ann. Appl. Biol. **49**, 316–330 (1961 b)

Lapwood, D. H.: Laboratory assessment of the susceptibility of potato haulm to blight (Phytophthora infestans). Eur. Potato J. **4**, 177–128 (1961 c)

Lauffer, M. A., Ansevin, A. T., Cartwright, T. E., Brinton, G. C.: Polymerization-depolymerization in tobacco mosaic virus protein. Nature (London) **181**, 1338–1339 (1958)

Law, I. J., Strijdom, B. W.: Some observations on plant lectins and Rhizobium specificity. Soil Biol. Biochem. **9**, 79–84 (1977)

Lelliott, R. A., Billing, E., Hayward, A. C.: A determinative scheme for fluorescent plant pathogenic bacteria. J. Appl. Bact. **29**, 470–489 (1966)

Leppik, E. E.: Gene centers of plants as sources of disease resistance. Ann. Rev. Phytopathol. **8**, 323–344 (1970)

Levine, M. N.: Biometric studies of the variation of physiologic forms of Puccinia graminis tritici and the effects of ecological factors on the susceptibility of wheat varieties. Phytopathology **18**, 7–123 (1928)

Lewis, D. H.: Concepts in fungal nutrition and the origin of biotrophy. Biol. Rev. **48**, 261–278 (1973)

Lim, S. M.: Diallel analysis for reaction of eight corn inbreds to Helminthosporium maydis race T. Phytopathology **65**, 10–15 (1975)

Lim, S. M., Kinsey, J. G., Hooker, A. L.: Inheritance of virulence in Helminthosporium turcicum to monogenic resistant corn. Phytopathology **64**, 1150–1151 (1974)

Lipetz, J.: Wound healing in higher plants. Int. Rev. Cytol. **27**, 1–28 (1970)

Littlefield, L. J.: Histological evidence for diverse mechanisms of resistance to flax rust, Melampsora lini (Ehrenb.) Lev. Physiol. Plant Pathol. **3**, 241–247 (1973)

Loegering, W. Q., Powers, H. R.: Inheritance of pathogenicity in a cross of physiological races 111 and 36 of Puccinia graminis f. sp. tritici. Phytopathology **52**, 547–554 (1962)

Luig, N. H., Rajaram, S.: The effect of temperature and genetic background on the host gene expression and interaction to Puccinia graminis tritici. Phytopathology **62**, 1171–1174 (1972)

Luig, N. H., Watson, I. A.: A study of inheritance of pathogenicity in Puccinia graminis var. tritici. Proc. Linn. Soc. N. S. W. **81**, 115–118 (1961)

Luig, N. H., Watson, I. A.: Studies on the genetic nature of resistance to Puccinia graminis var. tritici in six varieties of common wheat. Proc. Linn. Soc. N. S. W. **90**, 299–327 (1965)

Luke, H. H., Wheeler, H. E.: Toxin production by Helminthosporium victoriae. Phytopathology **45**, 453–458 (1955)

Lupton, F. G. H., Johnson, R.: Breeding for mature-plant resistance to yellow rust in wheat. Ann. Appl. Biol. **66**, 137–143 (1970)

Lyles, W. E., Futrell, M. C., Atkins, I. M.: Relation between reaction to race 15B of stem rust and reducing sugars and sucrose in wheat. Phytopathology **49**, 254–256 (1959)

Lyon, F., Wood, R. K. S.: The hypersensitive reaction and other responses of bean plants to bacteria. Ann. Bot. **40**, 479–491 (1976)

MacKenzie, D. R.: Application of two epidemiological models for the identification of slow stem rusting in wheat. Phytopathology **66**, 55–59 (1976)

Macko, V., Woodbury, W., Stahmann, M. A.: The effect of peroxidase on the germination and growth of mycelium of Puccinia graminis f. sp. tritici. Phytopathology **58**, 1250–1254 (1968)

Mahomet, H. A.: Temperature requirements for identification of races 49 and 139 of Puccinia graminis tritici. Phytopathology **44**, 498 (1954)

Mains, E. B.: The relation of some rusts to the physiology of their hosts. Am. J. Bot. **4**, 179–220 (1917)

Mains, E. B., Jackson, H. S.: Physiologic specialization in the leaf rust of wheat, Puccinia triticina Erikss. Phytopathology **16**, 89–120 (1926)

Martens, J. W.: Stem rust of oats in Canada in 1972. Can. Plant Dis. Surv. **52**, 171–172 (1972)

Martens, J. W.: Stem rust of oats in Canada in 1973. Can. Plant Dis. Surv. **54**, 19–20 (1974)

Martens, J. W.: Stem rust of oats in Canada in 1974. Can. Plant Dis. Surv. **55**, 61–62 (1975)

Martens, J. W., Anema, P. K.: Stem rust of oats in Canada in 1971. Can. Plant Dis. Surv. **52**, 17–18 (1972)

Martens, J. W., McKenzie, R. I. H.: Stem rust of oats in Canada in 1975. Can. Plant Dis. Surv. **56**, 23–24 (1976)

Martens, J. W., McKenzie, R. I. H., Green, G. J.: Thermal stability of stem rust resistance in oat seedlings. Can. J. Bot. **45**, 451–458 (1967)

Martin, T. J., Ellingboe, A. H.: Differences between compatible parasite/host genotypes involving the Pm4 locus of wheat and the corresponding genes in Erysiphe graminis f. sp. tritici. Phytopathology **66**, 1435–1438 (1976)

Mather, K.: Polygenic inheritance and natural selection. Biol. Rev. **18**, 32–64 (1943)

Matta, A.: Microbial penetration and immunization of uncongenial host plants. Ann. Rev. Phytopathol. **9**, 387–410 (1971)

McClure, M. A., Misaghi, I., Nigh, E. L.: Shared antigens of parasitic nematodes and host plants. Nature (London) **244**, 306–307 (1973)

McIntosh, R. A., Luig, N. H., Baker, E. P.: Genetic and cytogenic studies of stem rust, leaf rust, and powdery mildew resistances in Hope and related wheat cultivars. Aust. J. Biol. Sci. **20**, 1181–1192 (1967)

McKinney, H. H., Clayton, E. E.: Genotype and temperature in relation to symptoms caused in Nicotiana by the mosaic virus. J. Hered. **36**, 323–331 (1945)

Meehan, F., Murphy, H. C.: Differential phytotoxicity of metabolic by-products of Helminthosporium victoriae. Science **106**, 270–271 (1947)

Mercer, P. C., Wood, R. K. S., Greenwood, A. D.: Resistance to anthracnose of French bean. Physiol. Plant Pathol. **4**, 291–306 (1974)

Minamikawa, T., Uritani, I.: Phenylanaline ammonia-lyase in sliced sweet potato roots. J. Biochem. **57**, 678–688 (1965)

Morris, E. R., Rees, D. A., Young, G., Walkinshaw, M. D., Darke, A.: Orderdisorder transition of a bacterial polysaccharide in solution. A role for polysaccharide conformation in recognition between Xanthomonas pathogen and its plant host. J. Mol. Biol. **110**, 1–16 (1977)

Moscona, A.: Rotation-mediated histogenic aggregation of dissociated cells: A quantifiable approach to cell interactions in vitro. Exp. Cell. Res. **22**, 455–475 (1961)

Müller, K. O., Börger, H.: Experimentelle Untersuchungen über die Phytophthora-Resistenz der Kartoffel. Zugleich ein Beitrag zum Problem der „erworbenen Resistenz" im Pflanzenreich. Arb. Biol. Anst. (Reichanst.) Berlin **23**, 189–231 (1941)

Nelson, R. R.: Cochliobolus victoriae, the perfect stage of Helminthosporium victoriae. Phytopathology **50**, 774–775 (1960)

Newton, M., Brown, A. M.: Studies on the nature of disease resistance in cereals. I. The reactions to rust of mature and immature tissues. Can. J. Res. **11**, 564–581 (1934)

Newton, M., Johnson, T.: Stripe rust, Puccinia glumarum, in Canada. Can. J. Res. **14**, 98–108 (1936)

Newton, M., Johnson, T.: Experimental reaction of physiologic races of Puccinia triticina and their distribution in Canada. Can. J. Res. C **19**, 121–133 (1941)

Newton, M., Johnson, T.: Physiologic specialization of oat stem rust in Canada. Can. J. Res. C **22**, 201–216 (1944)

Nutman, P. S.: Genetics of symbiosis and nitrogen fixation in legumes. Proc. R. Soc. **172**, 417–437 (1969)

Ohno, S.: Evolution by Gene Duplication. Berlin-Heidelberg-New York: Springer 1970

Ohno, S.: Ancient linkage groups and frozen accidents. Nature (London) **244**, 259–262 (1973)

Oosawa, F., Asakura, S.: Thermodynamics of the Polymerization of Protein. London-New York-San Francisco: Academic 1975

Osoro, M. O., Green, G. J.: Stabilizing selection in Puccinia graminis tritici in Canada. Can. J. Bot. **54**, 2204–2214 (1976)

Otsuki, Y., Shimomura, T., Takebe, I.: Tobacco mosaic virus multiplication and expression of the N gene in necrotic responding tobacco varieties. Virology **50**, 45–50 (1972)

Parlevliet, J. E.: Partial resistance of barley to leaf rust, Puccinia hordei. III. The inheritance of the host plant effect on latent period in four cultivars. Euphytica **25**, 241–248 (1976a)

Parlevliet, J. E.: The genetics of seedling resistance to leaf rust, Puccinia hordei Otth. in some spring barley cultivars. Euphytica **25**, 249–254 (1976b)

Parlevliet, J. E.: Evaluation of the concept of horizontal resistance in the barley/Puccinia hordei host-pathogen relationship. Phytopathology **66**, 494–497 (1976c)

Parlevliet, J. E., Kuiper, H. J.: Resistance of some barley cultivars to leaf rust, Puccinia hordei; polygenic, partial resistance hidden by monogenic hypersensitivity. Neth. J. Plant Pathol. **83**, 85–89 (1977)

Parlevliet, J. E., Zadoks, J. C.: The integrated concept of disease resistance: a new view including horizontal and vertical resistance in plants. Euphytica **26**, 5–21 (1977)

Paxman, G. J.: Variation in Phytophthora infestans. Eur. Potato J. **6**, 14–23 (1963)

Pelham, J.: Resistance in tomato to tobacco mosaic virus. Euphytica **15**, 258–267 (1966)

Person, C., Sidhu, G.: Genetics of host-parasite interrelationships. In: Mutation Breeding for Disease Resistance, pp. 31–38. Vienna: International Atomic Energy Agency 1971

Peterman, M. A.: Relation of antigens in selected host-pathogen systems of Linum usitatissimum and Melampsora lini. M.S. thesis, North Dakota State University (1967). Quoted by Wimalajeewa and De Vay (1971)

Peturson, B.: Effect of temperature on host reactions to physiologic forms of Puccinia coronata avenae. Sci. Agric. **11**, 104–110 (1930)

Plaisted, R. L., Thurston, H. D., Tingey, W. M.: Five cycles of selection within a population of S. tuberosum spp. andigena. Am. Potato J. **52**, 280 (1975)

Pound, G. S., Cheo, P.-C.: Studies on resistance to cucumber mosaic virus 1 in spinach. Phytopathology **42**, 301–306 (1952)

Preston, R. D.: Versatile polysacchasides. Nature (London) **266**, 302–303 (1977)

Price, H. J.: Evolution of DNA content in higher plants. Bot. Rev. **42**, 27–51 (1976)

Pringle, R. B.: Chemistry of host-specific phytotoxins. In: Wood, R. K. S., Ballio, A., Graniti, A. (Eds.): Phytotoxins in Plant Diseases, pp. 139–155. London-New York: Academic 1972

Pringle, R. B., Scheffer, R. P.: Host-specific plant toxins. Ann. Rev. Phytopathol. **2**, 133–156 (1964)

Rajaram, S., Luig, N. H., Watson, I. A.: Inheritance of leaf rust resistance in four varieties of common wheat. Euphytica **20**, 574–585 (1971)

Reddi, K. K.: Studies on the formation of tobacco mosaic virus ribonucleic acid: II. Degradation of host ribonucleic acid following infection. Proc. Nat. Acad. Sci. U.S.A. **50**, 75–81 (1963)

Roberts, B. J., Moore, M. B.: The effects of temperature on the resistance of oat stem rust conditioned by the BC genes. Phytopathology **46**, 584 (1956)

Roberts, F. M.: The determination of physiologic forms of Puccinia triticina Eriksss. in England and Wales. Ann. Appl. Biol. **23**, 271–301 (1936)

Robinson, R. A.: Plant Pathosystems. Berlin-Heidelberg-New York: Springer 1976

Roelfs, A. P., McVey, D. V.: Races of Puccinia graminis f. sp. tritici in the U.S.A. during 1973. Plant Dis. Rep. **58**, 608–611 (1974)

Roelfs, A. P., McVey, D. V.: Races of Puccinia graminis f. sp. tritici in the U.S.A. during 1974. Plant Dis. Rep. **59**, 681–685 (1975)

Roelfs, A. P., McVey, D. V.: Races of Puccinia graminis f. sp. tritici in the U.S.A. during 1975. Plant Dis. Rep. **60**, 656–660 (1976)

Rohringer, R., Heitefuss, R.: Incorporation of P^{32} into ribonucleic acid of rusted wheat leaves. Can. J. Bot. **39**, 263–267 (1961)

Rohringer, R., Howes, N. K., Kim, W. K., Samborski, D. J.: Evidence for a gene-specific RNA determining resistance in wheat to stem rust. Nature (London) **249**, 585–587 (1974)

Roseman, S.: The synthesis of complex carbohydrates by multiglycosyl transferase systems and their potential functions in intercellular adhesion. Chem. Phys. Lipids **5**, 270–297 (1970)

Rosen, H. R.: New germ plasm for combined resistance to Helminthosporium blight and crown rust of oats. Phytopathology **45**, 219–221 (1955)

Rowell, J. B., Roelfs, A. P.: Evidence for an unrecognised source of overwintering wheat stem rust in the United States. Plant Dis. Rep. **55**, 990–992 (1971)

Rowley, D., Jenkin, C. R.: Antigenic cross-reaction between host and parasite as a possible cause of pathogenicity. Nature (London) **193**, 151–154 (1962)

Samborski, D. J.: Leaf rust of wheat in Canada in 1971. Can. Plant Dis. Surv. **52**, 8–10 (1972a)

Samborski, D. J.: Leaf rust of wheat in Canada in 1972. Can. Plant Dis. Surv. **52**, 168–170 (1972b)

Samborski, D. J.: Leaf rust of wheat in Canada in 1973. Can. Plant Dis. Surv. **54**, 8–10 (1974)

Samborski, D. J.: Leaf rust of wheat in Canada in 1974. Can. Plant Dis. Surv. **55**, 58–60 (1975)

Samborski, D. J.: Leaf rust of wheat in Canada in 1975. Can. Plant Dis. Surv. **56**, 12–14 (1976)

Samborski, D. J., Dyck, P. L.: Inheritance of virulence in wheat rust on the standard differential varieties. Can. J. Genet. Cytol. **10**, 24–32 (1968)

Samborski, D. J., Dyck, P. L.: Inheritance of virulence in Puccinia recondita in six backcross lines of wheat with single genes for resistance to leaf rust. Can. J. Bot. **54**, 1666–1671 (1976)

Samuel, G.: Some experiments on inoculation methods with plant viruses and on local lesions. Ann. Appl. Biol. **18**, 494–507 (1931)

Sanghi, A. K., Luig, N. H.: Resistance in three common wheat cultivars to Puccinia graminis. Euphytica **23**, 273–280 (1974)

Saxena, K. M. S., Hooker, A. L.: On the structure of a gene for resistance in maize. Proc. Nat. Acad. Sci. U.S.A. **61**, 1300–1305 (1968)

Scheffer, R. P.: A discussion. In: Wood, R. K. S., Graniti, A. (Eds.): Specificity in Plant Diseases, p. 100. New York-London: Plenum 1976

Scheffer, R. P., Yoder, O. C.: Host-specific toxins and selective toxicity. In: Wood, R. K. S., Ballio, A., Graniti, A. (Eds.): Phytotoxins in Plant Diseases, pp. 251–272. London-New York: Academic 1972

Schick, R., Möller, K. H., Haussdörfer, M., Schick, E.: Die Widerstandsfähigkeit von Kartoffelsorten gegenüber der durch Phytophthora infestans (Mont.) de Bary hervorgerufenen Krautfäule. Zuchter **28**, 99–105 (1958)

Schnathorst, W. C., De Vay, J. E.: Common antigens in Xanthomonas malvacearum and Gossypium hirsutum and their possible relationships to host specificity and disease resistance. Phytopathology **53**, 1143 (1963)

Schroeder, W. T., Provvidenti, R., Barton, D. W., Mishanec, W.: Temperature affecting a reversal of dominance in the resistance of Pisum sativum to bean virus 2. Phytopathology **50**, 654 (1960)

Schroeder, W. T., Provvidenti, R., Barton, D. W., Mischanec, W.: Temperature differentiation of genotypes for BV2 resistance in Pisum sativum. Phytopathology **56**, 113–117 (1966)

Seevers, P. M., Daly, J. M.: Studies on wheat stem rust resistance controlled by the Sr 6 locus. Phytopathology **60**, 1642–1647 (1970)

Seevers, P. M., Daly, J. M., Catedral, F. F.: The role of peroxidase isozymes in resistance to wheat stem rust disease. Plant Physiol. **48**, 353–360 (1971)

Sequeira, L., Graham, T. L.: Agglutination of avirulent strains of Pseudomonas solanacearum by potato lectins. Physiol. Plant Pathol. **11**, 43–54 (1977)

Sharp, E. L., Emge, R. G.: A "tissue transplant" technique for obtaining abundant sporulation of races of Puccinia graminis var. tritici on resistant varieties. Phytopathology **48**, 696–697 (1958)

Sharp, E. L., Sally, B. K., Taylor, G. A.: Incorporation of additive genes for stripe rust resistance in winter wheat. Phytopathology **66**, 74–80 (1976)

Shaw, C. R.: Electrophoretic variation in enzymes. Science **149**, 936–943 (1965)

Sidhu, G., Person, C.: Genetic control of virulence in Ustilago hordei. III. Identification of genes for host resistance and demonstration of gene-for-gene relations. Can. J. Genet. Cytol. **14**, 209–213 (1972)

Simmonds, N. W., Malcolmson, J. F.: Resistance to late blight in Andigena potatoes. Eur. Potato J. **10**, 161–166 (1967)

Simons, M. D.: The relationship of temperature and stage of growth to the crown rust reaction of certain varieties of oats. Phytopathology **44**, 221–223 (1954)

Singer, S. J., Nicolson, G. L.: The fluid mosaic model of the structure of cell membranes. Science **175**, 720–731 (1972)

Stakman, E. C., Fletcher, D. G.: The common barberry and black stem rust. U.S. Dep. Agric. Farmers' Bull. **1544** (1930)

Stakman, E. C., Lambert, E. B.: The relation of temperature during the growing season in the spring wheat area of the United States to the occurrence of stem rust epidemics. Phytopathology **18**, 369–374 (1928)

Stakman, E. C., Levine, M. N.: Effect of certain ecological factors on the morphology of the urediniospores of Puccinia graminis. J. Agric. Res. **16**, 43–77 (1919)

Steiner, G. W., Strobel, G. A.: Helminthosporoside, a host-specific toxin from Helminthosporium sacchari. J. Biol. Chem. **246**, 4350–4357 (1971)

Stevens, C. L., Lauffer, M. A.: Polymerization-depolymerization of tobacco mosaic protein. IV. The role of water. Biochemistry **4**, 31–37 (1965)

Stevenson, F. J., Akeley, R. V., Webb, R. E.: Reactions of potato varieties to late blight and insect injury as reflected in yields and total solids. Am. Potato J. **32**, 215–221 (1955)

Strobel, G. A.: The helminthosporiside-binding protein of sugarcane: Its properties and relationship to susceptibility to the eye spot disease. J. Biol. Chem. **248**, 1321–1328 (1973a)

Strobel, G. A.: Biochemical basis of the resistance of sugarcane to eyespot disease. Proc. Nat. Acad. Sci. U.S.A. **70**, 1693–1696 (1973b)

Strobel, G. A.: The toxin-binding protein of sugarcane, its role in the plant and in disease development. Proc. Nat. Acad. Sci. U.S.A. **71**, 4232–4236 (1974)

Strobel, G. A.: A mechanism of disease resistance in plants. Sci. Am. **232**, 80–88 (1975)

Strobel,G.A., Hapner,K.D.: Transfer of toxin susceptibility to plant protoplasts via the helminthosporiside binding protein of sugarcane. Biochem. Biophys. Res. Commun. **63**, 1151–1156 (1975)
Strobel,G.A., Hess,W.M.: Evidence for the presence of toxin-binding protein on the plasma membrane of sugarcane cells. Proc. Nat. Acad. Sci. U.S.A. **71**, 1413–1417 (1974)
Stroede,W.: Über den Einfluß von Temperatur und Licht auf die Keimung der Uredosporen von Puccinia glumarum f. sp. tritici (Schmidt) Erikss. et Henn. Phytopathol. Z. **5**, 613–624 (1933)
Tani,T., Yamamoto,H., Onoe,T., Naito,N.: Primary recognition and subsequent expression of resistance in oat leaves hypersensitively responding to crown rust fungus. In: Tomiyama,K., Daly,J.M., Uritani,I., Oku,H., Ouchi,S. (Eds.): Biochemistry and Cytology of Plant-Parasite Interaction, pp. 124–135. Tokyo: Kodansha 1976
Thom,M., Laetsch,W.M., Maretzki,A.: Isolation of membranes from sugarcane suspensions: evidence for a plasma membrane enriched fraction. Plant Sci. Lett. **5**, 245–253 (1975)
Thomas,H.R.: Factors affecting development of necrosis in some bean varieties inoculated with common bean mosaic virus. Phytopathology **44**, 508 (1954)
Thompson,J.N.: Quantitative variation and gene number. Nature (London) **258**, 665–668 (1975); **261**, 526 (1976)
Tomiyama,K.: Physiology and biochemistry of disease resistance of plants. Ann. Rev. Phytopathol. **1**, 295–324 (1963)
Tomiyama,K., Takakuwa,M., Takase,N.: The metabolic activity in healthy tissue neighbouring the infected cells in relation to resistance to Phytophthora infestans (Mont.) de By. in potatoes. Phytopathol. Z. **31**, 237–250 (1958)
Toms,E.C., Western,A.: Phytohaemagglutinins. In: Harborne,J.B., Boulter,D., Turner,B.L. (Eds.): Chemotaxonomy of the Leguminoseae, pp. 367–462. London-New York: Academic 1971
Toxopeus,H.J.: Reflections on the origin of new physiologic races of Phytophthora infestans and the breeding of resistance in potatoes. Euphytica **5**, 221–237 (1956)
Tu,J.C.: Cytoplasmic changes during and after infection of soybean root nodule cells with rhizobia. Phytopathology **66**, 1065–1071 (1976)
Uritani,I.: Protein changes in diseased plants. Ann. Rev. Phytopathol. **9**, 211–234 (1971)
Vanderplank,J.E.: Analysis of epidemics. In: Horsfall,J.G., Dimond,A.E. (Eds.): Plant Pathology, pp. 229–289. New York-London: Academic 1960
Vanderplank,J.E.: Plant Diseases: Epidemics and Control. New York-London: Academic 1963
Vanderplank,J.E.: Disease Resistance in Plants. New York-London: Academic 1968
Vanderplank,J.E.: Stability of resistance to Phytophthora infestans in cultivars without R genes. Potato Res. **14**, 263–270 (1971)
Vanderplank,J.E.: Principles of Plant Infection. New York-San Francisco-London: Academic 1975
Vanderplank,J.E.: Four essays. Ann. Rev. Phytopathology **14**, 1–10 (1976)
Vavilov,N.I.: The Origin, Variation, Immunity, and Breeding of Cultivated Plants. Waltham Massachusetts: Chronica Botanica 1949
Venkataraman,S., Takshminarasimham,C., Kalyanasundaram,R.: Antigenic determinants of host-pathogen specificity. Abst. 2nd Int. Congr. Plant Pathol. 0958 (1973)
Waldron,L.R.: Stem rust epidemics and wheat breeding. N. D. Agric. Exp. Stu. Circ. **57** (1935)
Ward,H.M.: On some relations between host and parasite in certain epidemic diseases of plants. Proc. R. Soc. London **47**, 393–443 (1890)
Ward,H.M.: On the relations between host and parasite in the bromes and their brown rust, Puccinia dispersa (Erikss.). Ann. Bot. **16**, 233–315 (1902)
Ward,H.M.: On the history of Uredo dispersa Erikss. and the "mycoplasm" hypothesis. Philos. Trans. R. Soc. London B **196**, 29–46 (1904)
Waterhouse,W.L.: Australian rust studies. Proc. Linn. Soc. N. S. W. **5**, 615–680 (1929)
Watson,I.A., Luig,N.H.: The classification of Puccinia graminis var. tritici in relation to breeding resistant varieties. Proc. Linn. Soc. N. S. W. **88**, 235–258 (1963)

Watson, I. A., Luig, N. H.: Sr 15—A new gene for use in the classification of Puccinia graminis var. tritici. Euphytica **15**, 239–250 (1966)

Webb, R. E., Bonde, R.: Physiological races of late blight fungus from potato dump-heap plants in Maine in 1955. Am. Potato J. **33**, 53–55 (1956)

Wells, H. D., Forbes, I.: Effect of temperature on growth of Glomerella cingulata in vitro and its pathogenicity to Lupinus angustifolius genotypes an an and An An. Phytopathology **57**, 1307–1311 (1967)

Wheeler, H.: Plant Pathogenesis. Berlin-Heidelberg-New York: Springer 1975

Wheeler, H.: Ultrastructure of penetration by Helminthosporium maydis. Physiol. Plant Pathol. **11**, 171–178 (1977)

Whitfield, H. J., Martin, R. G., Ames, B. N.: Classification of aminotransferase (C gene) mutants in the histidine operon. J. Mol. Biol. **21**, 335–355 (1966)

Whitney, H. S., Shaw, M., Naylor, J. M.: The physiology of host-pathogen relations. XII. A cytophotometric study of the distribution of DNA and RNA in rust-infected leaves. Can. J. Bot. **40**, 1533–1544 (1962)

Wilder, B. M., Albersheim, P.: The structure of plant cell walls. IV. A structural comparison of the wall hemicellulose of cell suspension cultures of sycamore (Acer pseudoplatanus) and of red kidney bean (Phaseolus vulgaris). Plant Physiol. **51**, 889–893 (1973)

Williams, N. D., Gough, F. J., Rondon, M. R.: Interaction of pathogenicitiy genes in Puccinia graminis f. sp. tritici and reaction genes in Triticum aestivum spp. vulgare "Marquis" and "Reliance." Crop Sci. **6**, 245–248 (1966)

Wimalajeewa, D. L. S., De Vay, J. E.: The occurence and characterization of a common antigen relationship between Ustilago maydis and Zea mays. Physiol. Plant Pathol. **1**, 523–535 (1971)

Winzler, R. J.: Carbohydrates in cell surfaces. Int. Rev. Cytol. **23**, 77–125 (1970)

Wolfe, M. S.: The genetics of barley mildew. Rev. Plant Pathol. **51**, 507–522 (1972)

Wolfe, M. S., Barrett, J. A., Shattock, R. C., Shaw, D. S., Whitbread, R.: Phenotype-phenotype analysis: field application of the gene-for-gene hypothesis in host-parasite relations. Ann. Appl. Biol. **82**, 369–374 (1976)

Yamamoto, M., Matsuo, K.: Involvement of DNA in resistance of potatoes to invasion by Phytophthora infestans. Nature (London) **259**, 63–64 (1976)

Yule, G. U.: On the theory of inheritance of quantitative compound characters on the basis of Mendel's laws—A preliminary note. Rep. 3rd Int. Conf. Genet. 140–142 (1906)

Zaag, D. E., van der: Overwintering en epidemiologie van Phytophthora infestans, tevens enige niewe bestrijdingsmogelijkheden. Tijdschr. Plantenziekten **62**, 89–156 (1956)

Zimmer, D. E., Schafer, J. F.: Relation of temperature to reaction type of Puccinia coronata on certain oat varieties. Phytopathology **51**, 202–203 (1961)

Subject Index

Aggressiveness of pathogens, definition 2
Amino acid residues 39
Amino acids
  with hydrophilic side-chains 38, 39, 41
  with hydrophobic side-chains 37, 38, 41
Antigens shared by host and pathogen 83—88
Apple *(Malus)* scab in a gene-for-gene system 22, 25
Association of virulence 106—108
Autoregulation 67, 68, 76
Avirulence
  definition 31
  primary role of genes 31, 32

Barley *(Hordeum)* resistance to brown rust 6—8, 22, 74
Bean *(Phaseolus)* resistance
  to mosaic virus 51
  to *Pseudomonas phaseolicola* 9
Berberis 32, 113, 115
Biotrophy 147—150
Botrytis cinerea 122
Brome mosaic virus RNA replicase 74

Ceratocystis fimbriata 84
C. ulmi 132
Cherry resistance to *Pseudomonas mors-prunorum* 5
Cladosporium fulvum in a gene-for-gene system 22, 25
Corn, see Maize
Cotton (*Gossypium* spp.)
  antigens shared with *Fusarium oxysporum* 87
  antigens shared with *Meloidogyne* 84
  antigens shared with *Verticillium albo-atrum* 87, 88
  resistance to *Xanthomonas malvacearum* 4, 5, 22, 26, 45, 51, 53, 55, 84, 97
Cronatium ribicola 132
Cucumber mosaic virus in *Spinacia* 51

Deletion of genes 28, 29, 32
Dissociation of virulence 105, 106
DNA 28, 30, 90, 91
Duplication (repetition) of genes 30, 31, 65, 150

Electrophoresis in incomplete analysis of isozymes 62, 63
Elicitors 94, 95
Endogone spp. 151
Erwinia amylovora 132
E. atroseptica 122
E. carotovora 122
Erwinia spp. 70, 122
Erysiphe graminis hordei 30
E. graminis tritici 22, 25, 51, 121, 122
Ethylene 72, 73

Flagellin 44, 45, 150
Flax *(Linum)*, resistance to *Melampsora lini* 9, 22, 31, 83—85
Fusarium caeruleum 99
F. oxysporum f. sp. *vasinfectum*, see Cotton

Gene-for-gene hypotheses
  Flor's 22, 23, 79—82, 140, 141
  second 111, 112
Gladiolus gandavensis 89
Glomerella cingulata 51, 78
Glycine max
  antigens common with *Meloidogyne incognita* 84
  elicitors 94, 95
  lectins 97, 98
Glycoproteins 92—96

Helminthosporium carbonum 136
H. maydis race T 128, 136, 138, 141, 149
H. sacchari 136, 138, 146
H. turcicum 80—82, 124, 125, 128, 130
H. victoriae 82, 136—138, 140—142, 144
Helminthosporoside 138, 139
Heterodera rostochiensis 22, 26
Hydrogen bonds
  in protein-protein recognition 41
  in RNA 91
  stabilizing protein structure 39, 44
Hydrophobic effect 35, 36, 41

Ipomoea batatas 84
Isopath lines 135
Isozymes 34, 35, 62, 63, 144

Lactate dehydrogenase 85
Lectins 97, 98
Leguminoseae 22, 97
Lilium longiflorum 43
Lupin *(Lupinus)* resistance to *Glomerella cingulata* 51, 78

Mahonia 32
Maize *(Zea mays)* 22, 30, 80—82, 84, 128—131, 136, 138, 141
Male-female recognition through pollen and stigma antigens 88, 89
Meloidogyne spp. 52, 84
Mitotic spindle, effect of temperature 43, 44
Multilines 133
Mutation
  forbidden 34
  missense 21, 63
  neutral 32—34, 144
  nonsense 21
  recurrent 32
  same-sense 21, 34, 37
  tolerable 34

Necrotrophy 147—149
Nonallelic virulence interaction
  hypothesis 109—111
  in *Puccinia graminis avenae* on genes Pg8 and Pg9 108
  in *P. graminis tritici* on genes Sr6, Sr9a, Sr9b, Sr9d and Sr9e 103—108
  in *P. recondita* on genes Lr17 and Lr18 108, 109
  in *P. striiformis* 108
Nucleus in host cell 23—27, 76, 146, 147

Oat (*Avena*), see *Helminthosporium victoriae*, *Ophiobolus graminis*, *Puccinia coronata*, *P. graminis avenae*, and *Ustilago maydis*
Operator-repressor model 67
Ophiobolus graminis, antigens shared with wheat and oats 88
Orobanche 22
Osmotic pressure in relation to susceptibility 59

Pathotoxins, selective 136—142
Pea (*Pisum*) resistance to bean yellow mosaic virus 51
Peptide linkage 39
Periconia circinata 136, 142
Peronospora tabacina 132
Peroxidase 66, 67, 69
Phenylanaline ammonia-lyase 67
Phyllosticta maydis 136
Phytoalexins 98, 99
Phytophthora megasperma var. *sojae* 94, 95
Phytophthora infestans, see also Gene-for-gene hypothesis, Nucleus, Resistance, horizontal, Resistance, vertical, Selection, directional, and Vertifolia effect 1—3, 8, 10—16, 22, 25, 27, 28, 33, 58, 71, 77, 99, 101, 111, 112, 121, 123, 128, 130—132, 134, 149
Plasmopara viticola 132
Polygenes, polygenic inheritance 126, 127
Polymerization of protein
  effect of solvent 57—59
  effect of temperature 43—45
Polyphenol oxidase 66—67
Polysaccharides, see Xanthans
Potato (*Solanum*), see *Heterodora rostochiensis*, *Phytophthora infestans*, *Synchytrium endobioticum*
Potato virus Y^n (tobacco veinal necrosis virus) 132, 133
Protein
  polymerization 35, 36
  recognition 40—42
  structure 39, 40
Protein-for-protein hypothesis (protein copolymerization hypothesis) 20ff.
Pseudomonas mori 70
P. mors-prunorum 5
P. phaseolicola 9, 70
P. solanacearum 98
P. syringae 70
Pseudoperonospora humuli 132
Puccinia coronata 24, 48, 60, 61, 72, 77, 137
P. graminis avenae 47, 48, 60, 108
P. graminis tritici, see also Gene-for-gene hypothesis, Nonallelic virulence interaction, Nucleus, Resistance, vertical, and Temperature 2—4, 9, 10, 21—25, 28, 29, 32, 33, 45—48, 59—61, 69, 70, 90, 102—108, 110—113, 134
P. hordei 6—8, 128, 129
P. polysora 24, 30, 124, 125, 128, 130—133
P. recondita 22, 24, 28, 29, 32, 45, 53—55, 60, 61, 64, 77, 108, 109
P. sorghi 22, 30, 77, 125, 128—131
P. striiformis 17—19, 22, 27, 52, 53, 57, 108, 125, 128, 129, 132

Qβ phage 74, 79
Quadratic check 82, 122, 140

Rank analysis 19
Resistance, adult (mature) plant 60, 61
Resistance genes, quality difference, see also Gene-for-gene hypotheses, second
  strong 112
  ultra weak 112
  weak, see also Mutation, neutral 112—114

Resistance, horizontal, see also Vertifolia effect
- accumulation 123—125
- anonymity of genes 129, 130
- in barley brown rust 128
- definition 2
- effect on epidemics 133—135
- gene numbers 128, 129, 142
- hypothesis 143—146
- loss of fitness 132, 133
- in maize leaf blight (northern) 124
- in maize leaf blight (southern) 128, 141, 142
- in maize rust caused by *Puccinia polysora* 124, 125
- in maize rust caused by *P. sorghi* 125, 126
- plague of false synonyms 120
- in potato blight 10—16, 123, 124
- specificity 130, 131
- tests 19
- with threshold 142
- in tomato blight 128
- in wheat stripe (yellow) rust 17—19, 125, 132

Resistance, vertical, see also Gene-for-gene hypotheses, Nonallelic virulence interactions, Selection, directional, Selection, stabilizing, Temperature, Vertifolia effect
- correlation of host and pathogen variation 9, 10
- definition 1, 2
- durability 117, 118
- effect on epidemics 8, 118, 119
- gene numbers 128, 129
- with host-pathogen specificity 2—4
- without host-pathogen specificity 4—8
- without hypersensitivity 9
- hypothesis 143—146
- loss of fitness 133
- in oat stem 108
- in potato blight 2, 3, 101
- tests 19
- use 114—117
- in wheat leaf rust 108, 109
- in wheat stem rust 2—4, 102—108
- in wheat stripe rust 108

Rhizobium spp. 22, 97, 98

RNA
- abundance 25, 148
- messenger 38, 91
- ribosomal 31, 75
- transfer 68

RNA replicase 73—75, 79

Sclerotium rolfsii 29, 149

Selection, directional 100, 101, 113—116

Selection, stabilizing 34, 102, 103, see also Nonallelic virulence interactions and Resistance genes, strong

Soft rot 70, 147

Sphaerotheca mors-uvae 132

Spinach (*Spinacia*), resistance to cucumber mosaic virus 51

Sorghum, see *Periconia circinata*

Sugarcane (*Saccharum*), selective pathotoxin 136, 138, 139

Susceptibility genes, primary role 27—30

Symbiosis 150, 151

Synchytrium endobioticum 22, 25, 26, 84

Temperature effect
- on dominance of resistance 50, 56, 57
- on protein monomers (denaturation) 44, 45
- on protein polymerization 43, 44
- on resistance in gene-for-gene systems 45—56
- on wheat stem rust epidemics 45, 46

Thucomyces lichenoides 150

Tilletia caries, *T. contraversa* 22

Tobacco mosaic virus
- in tobacco 51, 58, 59, 74
- in tomato 22, 49, 50, 57

Tobacco mosaic virus protein, effect of temperature 43

Tomato (*Lycopersicon*) 22, 48—50, 52, 57

Uncinula necator 132

Ustilago avenae 137

U. hordei 22, 25

U. maydis 84

Van der Waal's attraction 41

Variation
- erosion 144
- storage 21
- transformation from qualitative to quantitative 143—145

Vavilov's rule and corollary 29

Venturia inaequalis 22, 25

Verticillium albo-atrum 87, 88

Vertifolia effect 122, 131, 132

Victorin 137, 138

Virulence of pathogens, definition 1

Wheat (*Triticum*), see *Erysiphe graminis*, *Puccinia graminis*, *P. recondita*, *P. striiformis*, *Tilletia caries*, *T. contraversa*

Wounds 66, 67

Xanthans 96, 97

Xanthomonas malvacearum 4, 5, 22, 26, 45, 51, 53, 55, 84, 97

Advanced Series in Agricultural Sciences

Co-ordinating Editor:
B. Yaron
Editors: G. W. Thomas,
B. R. Sabey, Y. Vaadia,
L. D. Van Vleck

Volume 1: A. P. A. Vink
Land Use in Advancing Agriculture
1975. 94 figures, 115 tables. X, 394 pages
ISBN 3-540-07091-5

Contents: Land Use Surveys. – Land Utilization Types. – Land Resources. – Landscape Ecology and Land Conditions. – Land Evaluation. – Development of Land Use in Advancing Agriculture. – References. – Subject Index.

Volume 2: H. Wheeler
Plant Pathogenesis
1975. 19 figures, 5 tables. X, 106 pages
ISBN 3-540-07358-2

Contents: Concepts and Definitions. – Mechanisms of Pathogenesis. – Responses of Plants to Pathogens. – Disease-Resistance Mechanisms. – Genetics of Pathogenesis. – Nature of the Physiological Syndrome.

Volume 3: R. A. Robinson
Plant Pathosystems
1976. 15 figures, 2 tables. X, 184 pages
ISBN 3-540-07712-X

Contents: Systems. – Plant Pathosystems. – Vertical Pathosystem Analysis. – Vertical Pathosystem Management. – Horizontal Pathosystem Analysis. – Horizontal Pathosystem Management. – Polyphyletic Pathosystems. – Crop Vulnerability. – Conclusions. – Terminology.

Volume 4: H. C. Coppel, J. W. Mertins
Biological Insect Pest Suppression
1977. 46 figures, 1 table. XIII, 314 pages
ISBN 3-540-07931-9

Contents: Glossary. – Historical, Theoretical, and Philosophical Bases of Biological Insect Pest Suppression. – Organisms Used in Classical Biological Insect Pest Suppression. – Manipulation of the Biological Environment for Insect Pest Suppression. – A Fusion of Ideas. – References. – Index.

Volume 5: J. J. Hanan, W. D. Holley, K. L. Goldsberry
Greenhouse Management
1978. 283 figures, 150 tables. XIV, 530 pages
ISBN 3-540-08478-9

Contents: Light. – Greenhouse Construction. – Temperature. – Water. – Soils and Soil Mixtures. – Nutrition. – Carbon Dioxide and Pollution. – Insect and Disease Control. – Chemical Growth Regulation. – Business Management. – Marketing. – Appendices: Conversion Tables. Symbolism. Definitions.

Springer-Verlag
Berlin
Heidelberg
New York

Distribution rights for India:
Allied Publishers Private Ltd., New Delhi

Monographs on Theoretical and Applied Genetics

Co-ordinating Editor: R. Frankel
Editors: G.A.E. Gall, M. Grossman, H.F. Linskens, D. de Zeeuw

Volume 1
J. Sybenga
Meiotic Configurations
A Source of Information for Estimating Genetic Parameters
1975. 65 figures, 64 tables. X, 251 pages
ISBN 3-540-07347-7
Meiotic configurations are viewed from a special angle in this book: as a source for the extraction of maximum quantitative information of genetic interest, primarily related to recombination. This involves the development of models and systems for extimating genetic parameters from relatively simple microscopic observations on normal as well as specially constructed material. There are four chapters. Since information on chiasma formation is required for the effective analysis of chromosome pairing, after an introductory Chapter I, the analysis of crossing-over (Chapter 2), preceeds that on chromosome pairing (Chapter 3). Chromosome distribution (segregation) is treated in the final Chapter 4. Most of the material presented is based on published reports, some is new. Examples have been taken from various species, including plants and animals, and are not restricted to those with favorable chromosomes.

Springer-Verlag
Berlin
Heidelberg
New York

Volume 2
R. Frankel, E. Galun
Pollination Mechanisms, Reproduction and Plant Breeding
1977. 77 figures, 39 tables. XI, 281 pages
ISBN 3-540-07934-3
This book handles pollination mechanisms in a broad sense - i.e., the conveyance of the pollen grain to the female gametophyte, the germination and growth of the male gametophyte, fertilization, modes of pollen dispersal, the mechanisms assuring or preventing self-pollination, all in the light of plant breeding and crop production, so that the discussions are restricted to angiosperms and gymnosperms. The first part of the book describes modes of reproduction in higher plants and discusses ecology and dynamics of pollination; the second crops propagated by self-pollination and their specific breeding; the third details sexual reproduction in higher plants and the mechanisms involved in the prevention of self-pollination.
The book furnishes a broad fondation for crop production and plant breeding; it may also serve as a test in advanced plant breeding courses.

Volume 3
D. de Nettancourt
Incompatibility in Angiosperms
1977. 45. figures, 18 tables. XIII, 230 pages
ISBN 3-540-08112-7
This book classifies and discusses the enormous amount of isolated data accumulated on the physiology, biochemistry, ultrastructure, cytology, genetics and evolution of self-incompatibility and interspecific incompatibility in angiosperms. It outlays the probable evolution of self-incompatibility as a primitive outbreeding device and of interspecific incompatibility and self-compatibility as derived conditions. The various mechanisms of incompatibility known to operate in nature are presented in detail together with a critical examination of their importance for research in fundamental biology and in plant breeding. The numerous models and hypotheses elaborated to explain certain poorly understood features of incompatibility systems are reviewed. At the same time, and with special emphasis on the use of mutagens, chemicals and in vitro culture techniques, the author makes a detailed presentation of the arsenal of methods now available for bypassing incompatibility barriers and transforming the breeding behavior of higher plants.

CPSIA information can be obtained at www.ICGtesting.com
Printed in the USA
LVOW070225260912

300352LV00005B/12/P

9 783642 669675